D0450195

Praise for *The Lomborg Deception*

"*The Lomborg Deception*, by Howard Friel, presents a troubling history of how a cleverly contrived claim — that hundreds of scientists and dozens of scientific institutions have gotten climate and environmental science badly wrong over several decades — is way off base, unlike the well-established conclusions hammered out over decades in peer-reviewed assessments. Bjørn Lomborg's claims that environmental scientists mislead society into wasting money on nonexistent problems is based on hundreds of citations taken out of context, dozens of straw men, selective inattention to inconvenient science, and the illusion of careful scholarship. Friel documents this deception brilliantly.

The Lomborg Deception should serve as a sober warning to beware of the "myth busters and truth tellers" like Lomborg, who most likely are the ones misrepresenting complex environmental science problems — and, of course, profiting from the naive acceptance of seemingly careful claims that many wed to status quo policies so welcome."
— Stephen H. Schneider, Melvin and Joan Lane Professor for Interdisciplinary Environmental Studies and Senior Fellow, Woods Institute for the Environment, Stanford University

"For those interested in the future of polar bears and Arctic sea ice, *The Lomborg Deception*, by Howard Friel, clearly documents the inaccurate and utterly inadequate arguments that Bjørn Lomborg uses to erroneously suggest climate warming will have little negative effect on this bellwether mammal. The far greater tragedy is that misleading presentations such as those proffered by Lomborg may help to foster uncertainty in the public at large about the severity of the human causes of climate warming, and thus further delay the urgent need for the entire world to respond quickly to reduce our collective output of greenhouse gases."
— Ian Stirling, Fellow of the Royal Society of Canada, Research Scientist Emeritus, Environment

"For nearly a decade, Bjørn Lomborg's climate-science rejectionism has helped block serious political action on greenhouse emissions. We are now nearly out of time to act to prevent the worst impacts of global warming. I hope that Howard Friel's remarkably valuable and highly enlightening analysis in *The Lomborg Deception* will finally clear Lomborg and his supporters out of the path to such action."

—James Gustave Speth, author of *Red Sky at Morning: America and the Crisis of the Global Environment* and *The Bridge at the End of the World: Capitalism, the Environment, and the Crossing from Crisis to Sustainability*

"Facts, John Adams once said, are stubborn things. Unfortunately for Bjørn Lomborg, this book is full of them. *The Lomborg Deception* sets the record straight with a rigorous, readable body blow to climate complacency."

— Senator John Kerry

"Scientific discourse, born of the honest skepticism and unquestioned integrity of its participants, has been the lifeblood of scientific inquiry for centuries. False scientific debate over well-established results, born of the repeated interventions by voices whose skepticism is not honestly articulated, can be extremely dangerous — both to the conduct of scientific research and to the well-being of a planet that depends on responsible stewardship for its very existence. Since this danger is particularly acute for issues like climate change that have been politicized beyond reason, one can only hope that Howard Friel's careful documentation of persistent and pervasive misrepresentation will diminish significantly the credibility and thus the influence of Lomborg's assertions."

— Gary W. Yohe, Woodhouse/Sysco Professor of Economics, Wesleyan University

The Lomborg Deception

HOWARD FRIEL

The Lomborg Deception

Setting the Record Straight About Global Warming

FOREWORD BY THOMAS E. LOVEJOY

Yale
UNIVERSITY
PRESS
NEW HAVEN AND LONDON

Published with assistance from the Louis Stern
Memorial Fund.

Set in Janson type by Keystone Typesetting, Inc.
Printed in the United States of America.

Library of Congress Cataloging-in-Publication Data
Friel, Howard, 1955–
The Lomborg deception : setting the record straight about
global warming / Howard Friel.
p. cm.
Includes bibliographical references and index.
ISBN 978-0-300-16103-8 (hardcover : alk. paper)
1. Fraud in science. 2. Global warming. 3. Climatic
changes. 4. Lomborg, Bjørn, 1965. I. Title.
Q175.37.F75 2010
500—dc22
2009037999

A catalogue record for this book is available from the
British Library.

This paper meets the requirements of ANSI/NISO
Z39.48-1992 (Permanence of Paper).

10 9 8 7 6 5 4 3 2 1

CONTENTS

FOREWORD

THOMAS E. LOVEJOY

Ever a graduate student, I habitually turn to the back of a scholarly article or book to read the footnote or see what the citation actually is. Scholars do this not from some boring pedantic thoroughness, but rather out of true intellectual curiosity. As usual, I did that when reviewing the *Skeptical Environmentalist* for *Scientific American.* I remember my frustration at inadequate citations, so much so that I characterized them in the review as a "mirage in the desert." I reviewed only the forest and biodiversity aspects of the book as that was my particular expertise and assignment, and three others from different fields reviewed other aspects of the book. Little did I know that the entire volume was similarly flimsy.

I do recall at the time that fellow conservation biologists attending a Lomborg talk would correct his science, only to find the same assertions made in subsequent talks as if the corrections had never occurred. That left me disinclined to engage with Lomborg. Science and public understanding do not advance on the basis of assertions as opposed to conversations and discussion.

I do remember being puzzled at the time that Cambridge University Press had published the book, for surely a scholarly press would have picked up the problem in manuscript review. Later I came to understand that the review had been on the social science side of the Press even though the volume was in the Environmental Science list. So clearly there was a flaw in the reviewing process since it was an interdisciplinary subject. I still find it surprising the reviewer didn't question some of the assertions even if not an expert on environment from a scientific perspective.

Bjørn Lomborg was trained as an economist. Economics, of course, while contributing to environmental solutions in the case of some market based solutions like the sulfur markets in the United States, has some inherent difficulties in dealing with long term and big scale problems: witness the debate over discount rates between Sir Nicholas Stern and economists such as William Nordhaus. (The *Stern Report* excluded the use of discount rates because of the enormity and complicated time scales of climate change and its impacts.) The difference with the Lomborg approach is that Nordhaus is very rigorous: while using discount rates, he is impeccable about the understanding and use of physical and biological science.

In this work, Howard Friel does what nobody has done before, namely to systematically examine Lomborg's work citation by citation. This is no small task, so it is not surprising that this has not been undertaken until now.

Friel's work reveals the mirage to be pervasive, indeed, as big as the desert. This does not mean of, course, that everything that Lomborg writes is wrong or invalid but that it is a house of cards to a highly disturbing degree. Friel has used real scholarship to reveal the flimsy nature of the scholarly foundation of Lomborg's work.

What is unfortunate is that this took so long to come to light. A huge amount of time and energy has gone into addressing Lomborg's assertions, and the advance of policy about urgent environmental problems has been retarded.

The irony is that had Lomborg's scholarship been sound and some of the concerns of environmental scientists been demonstrated to have been incorrect, nobody would have been happier than scientists like myself. If there were not a grave and rapidly mushrooming biodiversity crisis, I could indulge in the intellectual joys of studying the marvels of life on Earth (my original motivation) without having to be concerned about biodiversity loss and ways to restrain it.

But environmental problems have indeed grown exponentially, with retreat of the Arctic ice, sea level rise now projected to rise a meter by century's end (and still probably underestimated), and the tipping point for dieback of the eastern half of the Amazon creeping closer. The rapidly changing global environment is beginning to seem like an Edgar Allan Poe short story.

In the meantime, the United States, once so much the global leader on environmental problems, is only now beginning to come to grips with environment and climate change. All along, like terriers nipping at heels, naysayers without the least qualifications delay and water down the process, and make the ultimate impact even greater for lack of strong and immediate action.

So let us hope a lesson has been learned, in particular that hope is false when based on poor scholarship. Even in this electronic age, where some students think any citation prior to the twenty-first century is irrelevant, and where it is so easy to troll for information in a cyber-world in which quality control is very uneven, it is still critical—with Kindle or whatever—to check the citation in the back of the book.

AUTHOR'S NOTE

The seemingly endemic problems in the writings of Bjørn Lomborg as they relate to global warming posed the additional challenge of coherently presenting those problems to the reader of this volume. It therefore seemed useful to identify the two main strains of arguments in his work as they pertain here. Thus, "Lomborg's Theorem" refers to his claim that anthropogenic (man-made) global warming is "no catastrophe." And "Lomborg's Corollary" represents his contention that since global warming is no catastrophe, there is little need to incur the costs of reducing greenhouse-gas emissions to the extent urged by concerned experts to avoid the worst impacts of global warming.

The focus of this volume is on Lomborg's Theorem as presented in his books *The Skeptical Environmentalist: Measuring the Real State of the World* (2001) and *Cool It: The Skeptical Environmentalist's Guide to Global Warming* (2007). The aim is to show that Lomborg's Theorem is grounded in highly questionable data and analysis, and that there is little if any factual or analytic basis for the theorem.

PART 1

Lomborg's Modus Operandi

one
2001: A THEOREM'S ODYSSEY

On September 10, 2001, Cambridge University Press published *The Skeptical Environmentalist: Measuring the Real State of the World*, by Bjørn Lomborg, a Danish statistician who argued that the "real" condition of the world's environment is better than what the major environmental organizations have routinely reported. Lomborg argued that the environmental groups—such as Greenpeace and the Worldwatch Institute—were too pessimistic and thus overstated humankind's harmful impact on the Earth's land, air, water, and animals.

At the outset, Lomborg maintained that there was little evidence to substantiate a gloomy picture of the Earth's environment. He rejected this view as the product of an exaggerated "Litany" of bad news generated by environmentalists:

> We are all familiar with the Litany: the environment is in poor shape here on Earth. Our resources are running out. The population is ever growing, leaving less and less to eat. The air and the water are becoming ever more polluted. The planet's species are becoming extinct is [*sic*] vast numbers—we kill off more than 40,000 each year. The forests are disappearing, fish stocks are collapsing and the coral reefs are dying.
>
> We are defiling our Earth, the fertile topsoil is disappearing, we are paving over nature, destroying the wilderness, decimating the biosphere, and will end up killing ourselves in the process. The world's ecosystem is breaking down. We are fast approaching the absolute limit of viability, and the limits of growth are becoming apparent.
>
> We know the Litany and have heard it so often that yet another repetition is, well, almost reassuring. There is just one problem: it does not seem to be backed up by the available evidence.[1]

Pursuant to these remarks—in 350 pages of text and nearly three thousand endnotes—Lomborg purportedly set out to expose the exaggerations of the environmentalists and uncover the underappreciated good news about the world's environment. Upon doing so, and writing heroically in the first person throughout his introductory remarks, Lomborg declared: "I will need to challenge our usual conception of the collapse of ecosystems, because this conception is simply not in keeping with reality."[2]

About global warming, Lomborg wrote that it is "almost certainly taking place," though its projected impact is "rather unrealistically pessimistic" and "will not pose a devastating problem for our future." About environmentalists' calls for a significant reduction of man-made greenhouse emissions, Lomborg argued that "the typical cure of early and radical fossil fuel cutbacks is way worse than the original affliction."[3]

From these and many similar statements, we can identify "Lomborg's Theorem," circa 2001, which asserts that the Earth and its environment are not threatened in any fundamental sense by human activity and, for the purposes of this volume, that man-made global warming is not the catastrophe that the environmental organizations claim. Lomborg's book, with its illusion of serious scholarship, given the number of endnotes, was influential in the United States throughout the presidential tenure of George W. Bush, who held power during a critically important window of opportunity to reduce greenhouse emissions to prevent the worst impacts of global warming. Probably more than any single published source, Lomborg's *The Skeptical Environmentalist* marked global warming as a threat that was "exaggerated" by environmentalists, and helped justify the inaction on greenhouse emissions by the Bush administration and the Republican-led Congress in the United States. Lomborg's influence was such that in 2004 *Time* named him one of the world's one hundred most influential people.[4]

By November 2007, Lomborg had updated his original analysis in *The Skeptical Environmentalist* with a book focused exclusively on climate change titled, *Cool It: The Skeptical Environmentalist's Guide to Global Warming*. He began this book as follows:

> Global warming has been portrayed recently as the greatest crisis in the history of civilization. As of this writing, stories on it occupy the front pages of *Time* and *Newsweek* and are featured prominently in countless media around the world. In the face of this level of unmitigated despair, it is perhaps surprising—and will by many be seen as inappropriate—to write a book that is basically optimistic about humanity's prospects.
>
> That humanity has caused a substantial rise in atmospheric carbon-dioxide levels over the past centuries, thereby contributing to global warming, is beyond debate. What is debatable, however, is whether hysteria and head-long spending on extravagant CO_2-cutting programs at an unprecedented price is the only possible response. Such a course is especially debatable in a world where billions of people live in poverty, where millions die of curable diseases, and where these lives could be saved, societies strengthened, and environments improved at a fraction of the cost.[5]

As in *The Skeptical Environmentalist*, in *Cool It* Lomborg doesn't doubt the phenomenon of human-induced warming. Rather, he argues that a warming of the Earth threatens "no catastrophe" for humanity or the Earth's environment; consequently, Lomborg sees no need to focus on significant reductions of greenhouse emissions as a matter of national or global policy.[6] The 2007 publication of *Cool It* thus updated and focused Lomborg's argument that the threat of climate change is exaggerated, and further reinforced the Bush administration's dissent from the scientific consensus on the need for major reductions in greenhouse emissions.

Lomborg's concession on one count—that global warming was happening and that it was predominately human-induced—conferred a superficial appearance of moderation between "the Litany" of the liberal environmentalists and the right-wing denials that CO_2 emissions were changing the Earth's climate. Lomborg emphasized this idea of a sober middle course by highlighting his conversion from left-wing environmental orthodoxy, noting in *The Skeptical Environmentalist* that "I'm an old left-wing Greenpeace member and had for a long time been concerned about environmental questions."[7] The concession and the conversion were inspired credentials from which to forge the "skeptical environmentalist" brand—used in the title of both books—and to invent a genre of anti-environmentalism for the ostensible benefit of the environment and humankind.

As one reads on, one might wonder how Lomborg's work managed to evade serious scrutiny by the major publishing houses—Cambridge University Press (2001) and Knopf (2007)—that issued his two major books, given Lomborg's problematic scholarship (as this volume will detail), and the importance of the global environmental issues that he addressed. Though *The Skeptical Environmentalist* declares at the outset that it "is critical of the way in which many environmental organizations make selective and misleading use of the scientific evidence,"[8] thus emphasizing its scientific implications, as Stephen Schneider noted, it "was published by the social science side of the house" at Cambridge University Press. Schneider, a prominent climate scientist, wrote that it was thus "not surprising that the [in-house] reviewers failed to spot Lomborg's unbalanced presentation of the natural science, given the complexity of the many intertwining fields."[9]

As *The Skeptical Environmentalist* progressed from its physical creation at Cambridge University Press to book reviews in major newspapers and journals, it somehow survived that level of scrutiny as well. Nicholas Wade, a veteran science editor and writer for the *New York Times*, seemed favorably disposed to Lomborg's environmental optimism. In one of the earliest incantations of the news media's repetitious descriptions of Lomborg's alleged environmental epiphany, Wade described him as "a vegetarian, backpack-toting academic who was a member of Greenpeace for four years," and acclaimed the "substantial work of analysis with almost 3,000 footnotes."[10]

In its review, the *Washington Post* depicted the skeptical environmentalist (the person) as "a self-described left-winger and former Greenpeace member" who "feels at one with the basic sentiments that underlie the Green movement." Lomborg is a "vegetarian with ethical objections to eating flesh" who wrote "a massive, meticulously presented argument that extends over 500 pages, supported by nearly

3,000 footnotes and 182 tables and diagrams," and who "found on close analysis that the factual foundation on which the environmental doomsayers stood was deeply flawed." This review in the *Post* found that *The Skeptical Environmentalist* (the book) demonstrates "emphatically" that "the population bomb is fizzling, and, far from killing us, pesticides and chemicals are improving longevity and the quality of life."[11]

Like the *Times* and the *Post*, the *Wall Street Journal*'s review observed that Lomborg's *The Skeptical Environmentalist* is "a superbly documented and readable book by a former member of Greenpeace" and "a self-described 'man of the Left.'" "Using uncontroversial data," the *Journal* continued, "Mr. Lomborg shows that the environment is improving, and the state of humanity too." And "as for global warming, Mr. Lomborg shows that it is unlikely to be catastrophic," and "even if temperatures increase substantially, Mr. Lomborg argues, a draconian cut in fossil-fuel use is not the answer."[12]

Whereas Lomborg was favorably reviewed in the three most important newspapers in the United States, he was challenged more rigorously by the scientific and environmentalist communities that were the critical subjects of his book. Shortly after *The Skeptical Environmentalist* was published, at least three scientific forums were organized to respond to Lomborg's analysis. One such forum was posted in December 2001 on Grist, a Web site of "environmental news and commentary,"[13] where several commentators were invited to submit responses to Lomborg. These included: Lester Brown, founder and president of the Earth Policy Institute and founder and former president of the Worldwatch Institute; the Harvard biologist Edward O. Wilson; Norman Myers, a prominent and prolific scientist on biodiversity and species extinction; and Stanford University scientist Stephen Schneider, who is lead author and coauthor of a number of chapters in the major assessment reports on global warming by the UN's Intergovernmental Panel on Climate Change (IPCC).[14]

As founder and former president of the Worldwatch Institute, Brown was the senior author of the institute's annual *State of the World* reports, which detailed environmental problems worldwide. The subtitle of Lomborg's book—*The Skeptical Environmentalist: Measuring the Real State of the World*—intentionally co-opts the name of these reports. Brown began his response to Lomborg: "Some years ago, well before many outside Denmark knew of Bjørn Lomborg's name, a group of his fellow faculty members at the University of Aarhus took the unusual step of developing a website specifically to warn the scientific community and others about flaws in his work. Appalled by Lomborg's scientific pretensions and unfounded conclusions, these faculty members, including a former head of the Danish Academy of Sciences, actively disassociated themselves from him. . . . Lomborg's fellow faculty members are concerned that his work does not satisfy basic academic standards."[15]

Continuing, Brown observed that Lomborg's thesis "is that the environmental movement has overstated the magnitude of environmental threats." Brown then noted that "a serious test of this hypothesis would require a systematic review of the research output of the leading environmental groups, tabulating both the instances where they have overstated and where they have understated threats to the environment." Upon noting other prerequisites that, in Brown's mind, Lomborg did not meet—including "determining which threats identified by environmental groups turned out to be real and which did not," and "tabulating those issues that environmentalists either missed entirely or identified only belatedly"—Brown argued that "only with such an approach could one decide whether environmentalists as a group have overstated or understated the threats to our planet." Brown concluded: "In failing to take such an approach, Lomborg's book becomes nothing more than a diatribe."[16]

Wilson responded a bit more pointedly: "My greatest regret about the Lomborg scam is the extraordinary amount of scientific talent that has to be expended to combat it in the media." Wilson described Lomborg's book as "characterized by willful ignorance, selective quotations, disregard for communication with genuine experts, and destructive campaigning to attract the attention of the media rather than scientists." Referring specifically to Lomborg's claim that environmentalists have exaggerated rates of species extinction, Wilson wrote that "Lomborg's estimate of extinction rates is at odds with the vast majority of respected scholarship on extinction," and, "at current levels of habitat destruction, extinction rates are destined to rise, and—I believe every researcher would agree—dramatically so."[17]

Myers, who debated Lomborg's mentor, Julian Simon, in 1992 at Columbia University, also responded to Lomborg over the issue of species extinction.[18] Like Brown and Wilson, Myers found serious problems with Lomborg's methods: "Bjorn Lomborg opens his chapter on biodiversity by citing my 1979 estimate of 40,000 species lost per year. He gets a lot of mileage out of that estimate throughout the chapter, although he does not cite any of my subsequent writings except for a single mention of a 1983 paper and a 1999 paper, neither of which deals much with extinction rates. Why doesn't he refer to the 80-plus papers I have published on biodiversity and mass extinction during the 20-year interim? In this respect, as well as others, Lomborg seems to be exceptionally selective."[19]

According to Myers: "Lomborg is equally sloppy in his analyses of the utilitarian benefits of species and their genetic resources"; "Lomborg seems disinclined to undertake even a fraction of the homework that could give him a preliminary understanding of the science in question [biodiversity and species extinction]"; and "Lomborg ignores or is ignorant of much of the work on extinction rates."[20]

Echoing his colleagues' complaints about Lomborg's methodology, Schneider focused on Lomborg's analysis of global warming in *The Skeptical Environmentalist:* "Bjorn Lomborg's chapter on global climate change is a clever polemic; it seems like a sober and well-researched presentation of balanced information, whereas in fact it makes use of selective inattention to inconvenient literature and overemphasis of work that supports his lopsided views. The Intergovernmental Panel on Climate Change reports and other honest assessments don't have the luxury of using such tactics, given the hundreds of external reviewers and dozens of review editors."[21]

In a section of his response to Lomborg titled, "On the Media," Schneider continued: "The real travesty is that the mainstream media have quoted *The Skeptical Environmentalist* as if it contained something new—some original analysis the rest of the community had missed, or some more balanced assessment. The sooner Lomborg's own unbalanced and incomplete 'analysis' is exposed, the better we will all be." Schneider further objected to "scores upon scores of strawmen, misquotes, unbalanced statements, and selective inattention to the full literature," in addition to Lomborg's "flimsy Greenpeace connection."[22]

Another such forum was initiated shortly after the publication of *The Skeptical Environmentalist* by the Union of Concerned Scientists (UCS), which is based in Cambridge, Massachusetts. The forum's participants, leading scientists in their fields, were Peter Gleick (an expert on freshwater resources), Jerry Mahlman (an atmospheric scientist and climate modeler), Edward O. Wilson, Thomas Lovejoy (at the time the World Bank's chief biodiversity adviser), Norman Myers, Jeffrey Harvey (a physicist at the University of Chicago), and Stuart Pimm (a professor of biodiversity and conservation biology at Duke University).

UCS introduced the forum with a background statement summarizing Lomborg's claims that "population growth is not a problem, that there is plenty of freshwater around, that deforestation rates and species extinctions are grossly exaggerated, that the pollution battle has been won, and that global warming is too expensive to fix." The introductory comments by UCS also noted that "the heavily promoted book [*The Skeptical Environmentalist*], published by Cambridge University Press, has received significant attention from the media and praise from commentators writing in the *Economist, New York Times*, and *Washington Post.*" UCS then asked: "Does this book merit such positive attention? Does Lomborg provide new insights? Are his claims supported by the data?"[23]

UCS answered that the separately contributed reviews to its forum "unequivocally demonstrate that on closer inspection, Lomborg's book is seriously flawed and fails to meet basic standards of credible scientific analysis," and that "Lomborg consistently misuses, misrepresents or misinterprets data to greatly underestimate rates of species extinction, ignore evidence that billions of people lack access to clean water and sanitation, and minimize the extent and impacts of global warming due to the burning of fossil fuels and other human-caused emissions of heat-trapping gases."[24]

A third authoritative response to the publication of *The Skeptical Environmentalist* was published in January 2002 in *Scientific American.* As in the previous two forums, the commentators were distinguished scientists: Stephen Schneider, John P. Holdren (a chaired professor of environmental policy and science at Harvard University), John Bongaarts (former chair of the Panel on Population Projections at the National Research Council of the National Academy of Sciences), and Thomas Lovejoy (Biodiversity Chair at the Heinz Center).[25]

Schneider began his response by summarizing four major arguments from Lomborg's *The Skeptical Environmentalist:* climate science is uncertain; greenhouse emissions and average global temperatures will increase at or below the IPCC's lowest estimates; the benefits of a major global effort to mitigate the effects of global warming by reducing greenhouse emissions would not be worth the cost to the global economy; and the Kyoto Protocol to the 1992 United Nations Framework Convention on Climate Change (UNFCCC) is too expensive to implement and would only slightly reduce greenhouse emissions and global temperatures by the end of the twenty-first century.

Responding to the first point—the science of global warming is too uncertain to make long-term projections about the Earth's climate—Schneider noted that, to support this assertion, Lomborg depends on a "controversial" climate theory by Massachusetts Institute of Technology meteorologist Richard Lindzen, which, if accepted by the IPCC, would reduce the climate sensitivity range to human-induced greenhouse emissions by a factor of three. Schneider noted that Lomborg "fails either to understand [Lindzen's] mechanism or to tell us that it is based on only a few years of data in a small part of one ocean." Schneider, who, like Lindzen, is an expert on clouds and their effects on the Earth's climate, observed that Lomborg also cited "a controversial hypothesis from Danish cloud physicists" to provide "an alternative to carbon dioxide for explaining recent climate change." After briefly noting the technical aspects of the Danish hypothesis, Schneider commented that "the IPCC discounts this theory" because its power to explain global warming is not "sufficient to match that of much more parsimonious theories, such as anthropogenic [man-made] forcing."[26]

Regarding Lomborg's claim that greenhouse emissions and average global temperatures over the course of this century will be as low or lower than the lowest IPCC estimate,[27] Schneider wrote that "Lomborg asserts that over the next several decades new, improved solar machines and other renewable technologies will crowd fossil fuels off the market," and "this will be done so efficiently" that "the IPCC scenarios vastly overestimate" increases in atmospheric carbon dioxide over the same period. Schneider observed, however, that Lomborg cited only one study to support this scenario while ignoring that the economists he generally relies upon "strongly believe high emissions are quite likely." Schneider also argued that "Lomborg's most egregious distortions and poorest analyses are his citations of cost-benefit calculations" with respect to the relative costs and benefits of reducing greenhouse emissions. He responded to Lomborg's claim that the Kyoto Protocol is too expensive to implement, and would only negligibly reduce greenhouse emissions by year 2100, by noting that Lomborg confuses Kyoto's "decade-long protocol" for reducing greenhouse emissions for "a 100-year regime." Schneider noted that "Kyoto is a starting point" and "yet Lomborg, with his creation of a straw-man 100-year projection, would squash even this first step."[28]

As a participant in the *Scientific American* forum on *The Skeptical Environmentalist*, John P. Holdren, who currently is chief science adviser to President Barack Obama, wrote that Lomborg's chapter on energy "is devoted almost entirely to attacking the belief that the world is running out of energy, a belief Lomborg appears to regard as part of the 'environmental litany' but that few if any environmentalists actually hold." Holdren continued: "What environmentalists mainly say on this topic is not that we are running out of energy but that we are running out of environment—that is, running out of the capacity of air, water, soil and biota to absorb, without intolerable consequences for human well-being, the effects of energy extraction, transport, transformation and use."[29]

Responding to Lomborg's notion that human population growth is not a significant environmental problem, John Bongaarts wrote that Lomborg is "simply wrong" to argue that "the number of people is not the problem." Commenting, like his colleagues, on Lomborg's methodology, Bongaarts wrote:

> Past population growth has led to high population densities in many countries. Lomborg dismisses concerns about this issue based on a simplistic and misleading calculation of density as the ratio of people to all land. Clearly, a more useful and accurate indicator of density would be based on the land that remains after excluding areas unsuited for human habitation or agriculture, such as deserts and inaccessible mountains. For example, according to his simple calculation, the population density of Egypt equals a manageable 68 persons per square kilometer, but if the unirrigated Egyptian deserts are excluded, density is an extraordinary 2,000 per square kilometer. . . . Measured properly, population densities have reached extremely high levels, particularly in large countries in Asia and the Middle East.

Lomborg argues that poverty, not population, is the main cause of hunger and malnutrition. Bongaarts responded: "Lomborg correctly notes that poverty is the main cause of hunger and malnutrition, but he neglects the contribution of population growth to poverty." Bongaarts continued: "Lomborg approvingly notes the huge ongoing migration from villages to cities in the developing world. This has been considered a welcome development, because urban dwellers generally have higher standards of living than villagers. Because the flow of migrants is now so large, however, it tends to overwhelm the absorptive capacity of cities, and many migrants end up living in appalling conditions in slums. The traditional urban advantage is eroding in the poorest countries, and the health conditions in slums are often as adverse as in rural areas."[30]

Bongaarts concluded: "Population is not the main cause of the world's social, economic and environmental problems, but it contributes substantially to many of them. If population had grown less rapidly in the past, we would be better off now. And if future growth can be slowed, future generations will be better off."[31]

Another authoritative contributor to the *Scientific American* forum on Lomborg's book was Thomas Lovejoy. Like his colleagues, Lovejoy noted Lomborg's problematic assumptions, arguments, and conclusions. For example, Lovejoy, a leading conservation biologist, remarked that it was "disconcerting" to find that "Lomborg begins the chapter on biodiversity with a section questioning whether biodiversity is important." While Lomborg argues that environmentalists overstate the degree to which species are becoming extinct, Lovejoy noted: "When [Lomborg] finally gets to extinction, he totally confounds the process by which a species is judged to be extinct with the estimates and projections of extinction rates." More specifically, Lovejoy wrote: "Estimates of present extinction rates range from 100 to 1,000 times normal, with most estimates at 1,000. The percent of bird (12), mammal (18), fish (5) and flowering plant (8) species threatened with extinction is consistent with that estimate. And the rates are certain to rise—and to do so exponentially—as natural habitats continue to dwindle."[32]

Lovejoy noted other concerns with Lomborg's assertions about species extinction. For example, in a section in *The Skeptical Environmentalist* titled, "Models and Reality," Lomborg questioned, in his words, the "appealingly intuitive" scientific model that ties the stability and diversity of species to habitat size. "Its logic is," Lomborg wrote, "that the more space there is, the more species can exist." Seeking to undermine this established assumption, Lomborg wrote that the eastern forests of the United States "were reduced over two centuries to fragments totaling just 1–2 percent of their original area, but nonetheless this resulted in the extinction of only one forest bird." Lomborg argued that this and other examples highlight "a serious problem with [E. O.] Wilson's rule of thumb" that links loss of species to reduced habitat areas. To Lomborg's claim that only one species of bird suffered extinction when the eastern U.S. forests were reduced to 1–2 percent of their original area, Lovejoy responded that "only the old-growth forests shrank that much; total forest cover never fell below roughly 50 percent—allowing much biodiversity to survive as forest returned to an even greater area." Thus, Lomborg's scenario "does not contradict what species-area considerations predict but instead confirms them."[33]

Lovejoy also wrote: "[Lomborg's] consideration of acid rain in a separate chapter is equally poorly researched and presented. Indeed, the research is so shallow that almost no citation from the peer-reviewed literature appears. Lomborg asserts that big-city pollution has nothing to do with acid rain, when it is fact that nitrogen compounds (NOx) from traffic are a major source." Referring also to what he described as a pattern of "denial" in *The Skeptical Environmentalist* about global environmental realities, Lovejoy wrote:

> The pattern is evident in the selective quoting. In trying to show that it is impossible to establish the extinction rate, he states: "Colinvaux admits in *Scientific American* that the rate is 'incalculable,'" when Paul A. Colinvaux's text, published in May 1989, is: "As human beings lay waste to massive tracts of vegetation, an incalculable and unprecedented number of species are rapidly becoming extinct." Why not show that Colinvaux thought the number [of extinctions] is large? Biased language, such as "admits" in this instance, permeates the book.

Along similar lines, Lovejoy also wrote: "In addition to errors of bias, [Lomborg's] text is rife with careless mistakes. Time and again I sought to track references from the text to the footnotes to the bibliography to find but a mirage in the desert."[34]

In summary, and upon considering the Grist, Union of Concerned Scientists, and the *Scientific American* forums, it is difficult to recall a book by a major academic publisher that has engendered as much criticism from such a solid line of distinguished scientists. It is also worth pointing out an obvious fact—that every book contains mistakes. But when such mistakes occur "time and again" while reflecting a consistent didactic orientation, such "mistakes" may require another label as to what they are.

Lovejoy also criticized Lomborg's analysis of the state of the world's forests. Lomborg wrote that "there are no grounds for making such claims" about disappearing forests worldwide.[35] To support this point, Lomborg wrote at length about how environmentalists, in his view, had exaggerated the damage of the 1997–98 forest fires in Indonesia:

> Finally, we heard a great deal about the forest fires in Indonesia in 1997, which for months laid a thick layer of smog over all of Southeast Asia from Thailand to the Philippines. The fires constituted a genuine health problem and with a total cost of almost 2 percent of GDP had appreciable economic impact. However, they were also exploited as a means to focus attention on deforestation. The WWF [the Worldwide Fund for Nature] proclaimed 1997 as "the year the world caught fire" and their president, Claude Martin, stated unequivocally that "this is not just an emergency, it is a planetary disaster." Summing up, WWF maintained that, "in 1997, fire burned more forests than at any other time in history."
>
> This is not the case, however. In their report, WWF estimated that the fires in Indonesia involved 2 million hectares, despite the fact that this is *higher* than any other estimate cited in the report. Although the 2 million hectares are mentioned constantly, it is only well into the text that it becomes apparent that the figure comprises both forest and "non-forest" areas. The official Indonesian estimate was about 165,000–219,000 hectares. Later, satellite-aided counting has indicated that upwards of 1.3 million hectares of forests and timber areas may have burnt. The independent fire expert Johann Goldammer said that "there is no indication at all that 1997 was an extraordinary fire year for Indonesia or the world at large."[36]

Lovejoy responded: "Lomborg's discussion of the great fire in Indonesia in 1997 is still another instance of misleading readers with selective information. Yes, the WWF first estimated the amount of forest burned at two million hectares, and Indonesia countered with official estimates of 165,000 to 219,000 hectares. But Lomborg fails to mention that the latter [Indonesia's estimates] were not in the least credible and that in 1999 the Indonesian government and donor agencies, including the World Bank, signed off on a report that the real number was 4.6 million hectares."[37]

In addition, while Lomborg argued that the WWF was wrong to observe that fire burned more forests in 1997 than in any other year, Lomborg neglected to mention that the WWF report to which he referred cited several other major forest fires that year, as indicated in the first sentence of the WWF report: "In 1997, vast forest fires in Indonesia, Papua New Guinea, Brazil, Colombia and Africa focused attention onto what is rapidly becoming a global crisis."[38] While mentioning only the fires in Indonesia, Lomborg accused WWF president Claude Martin of "exploiting" those fires by calling them "a planetary disaster," when in fact Martin issued that characterization immediately following references, in his words, "to other fires—in Africa, Asia, the Americas, Europe and the Pacific—where the tragic events are being duplicated in many other forest ecosystems."[39] Lomborg also claimed that "the independent fire expert Johann Goldammer said that 'there is no indication at all that 1997 was an extraordinary fire year for Indonesia or the world at large,'" yet Goldammer's conclusion to this effect, according to Lomborg's endnote, was provided as a "personal communication."[40]

Contributing to the Grist forum on Lomborg, Emily Matthews, a senior associate at the World Resources Institute, also commented on Lomborg's analysis on forests: "In *The Skeptical Environmentalist*, Bjorn Lomborg writes that 'basically, the world's forests are not under threat.' A charitable reader could attribute this flawed conclusion to errors of omission and ignorance; perhaps the author simply doesn't know the sources well enough to interpret them properly. Less charitably, one might reasonably conclude that Lomborg intentionally selects his data and citations to distort or even reverse the truth. His interpretations of data on global forest cover and Indonesian forest fires aptly illustrate both failings."[41]

Though many readers might find the analysis above from several highly qualified commentators to be disturbing, I found them, ironically, reassuring, since I had independently encountered similar problems in Lomborg's 2007 book *Cool It*, and thus sought to learn whether others had previously expressed similar concerns to my own at the time.

My path to Lomborg did not begin with *The Skeptical Environmentalist* or the forums on Lomborg sponsored by Grist, the Union of Concerned Scientists, and *Scientific American*. It began as a book that I was planning to write about how the *New York Times* and *Wall Street Journal* had covered global warming over the past two decades. I had written (with Richard Falk) two other books about the *Times*'s coverage of major foreign policy issues,[42] and had planned a third volume on global warming. Those plans began to change after reading Lomborg's 2007 book, *Cool It*, which roughly coincided in fall 2007 with the release of the IPCC's synthesis report issued in November 2007,[43] the Nobel Prize lectures by Al Gore and IPCC chair Rajendra K. Pachauri in Oslo, Norway, in December 2007,[44] and news reports in January 2008 of the death of Bert Bolin, who was influential in the establishment of the IPCC and who served as its chair from 1988 to 1997.[45] And relative to other books that I had read at the time (including *Field Notes from a Catastrophe*, by Elizabeth Kolbert;[46] *Red Sky at Morning*, by Gus Speth;[47] and *The Discovery of Global Warming*, by Spencer Weart[48]—each one superb and distinctively important), Lomborg's *Cool It*, even on a prima facie basis, in addition to being intellectually unconvincing, was aesthetically unsettling, given his woolly locutions (as in so and so "tells us" or tells a "story," thereby inferring but not proving an environmentalist's exaggeration) and the swampy document referencing (as I point out below).

In reporting Bolin's death, the *New York Times* noted that he had traveled to Washington in 1959 to alert the National Academy of Sciences about human-induced global warming.[49] It was gratifying to read that the eighty-two-year-old Bolin had lived to see the IPCC awarded the Nobel Prize, noting that he "was thrilled," though it seemed a bitter irony that he would die less than a month later with apparently less influence in the White House and Congress—nearly fifty years after his trip to Washington — than Bjørn Lomborg, whose analysis of the threat of global warming had prevailed in recent years among the highest government officials in the United States.

As a preliminary matter, it should take only a few pages to compare Lomborg's major claims in *Cool It* — that the "consequences of global warming are often wildly exaggerated," and that "large and very expensive CO_2 cuts made now will have only a rather small and insignificant impact far into the future"[50] — to the prevailing scientific consensus on these issues throughout the past few decades. But briefly challenging Lomborg's claims will not fully explain the Lomborg phenomenon, which combines a near total absence of methodological integrity with outsized political influence. Thus, to weaken forthwith (a) Lomborg's Theorem (that global warming is "no catastrophe"), and (b) Lomborg's Corollary (that we therefore should not prioritize the reduction of greenhouse emissions), is to pull the ribbon that will begin the unraveling of Lomborg's alleged scholarship. To this end, a good beginning would be to situate Lomborg's assessment of the threat of global warming, and his advice on what to do about it, in the context of expert scientific opinion going back nearly forty years.

In 1971 a panel of climate experts from fourteen nations convened in Stockholm, Sweden. Spencer Weart, a science historian at Harvard University, noted, "It was the first major conference to focus entirely on a 'Study of Man's Impact on Climate.'" About the conference, Weart wrote: "Exhaustive discussions brought no consensus on what was likely to happen, but all agreed that serious changes were possible. The widely read report concluded with a ringing call for attention to the dangers of humanity's emissions of particle pollutants and greenhouse gases. The climate could shift dangerously 'in the next hundred years,' the scientists declared, 'as a result of man's activities.'"[51]

In 1979 the U.S. National Academy of Sciences issued a report on climate change that had been requested by President Jimmy Carter. Chaired by Jule Charney, a meteorologist at the Massachusetts Institute of Technology, the Ad Hoc Study Group on Carbon Dioxide and Climate, which consisted of several top scientists,[52] concluded:

"We now have incontrovertible evidence that the atmosphere is indeed changing and that we ourselves contribute to that change. Atmospheric concentrations of carbon dioxide are steadily increasing, and these changes are linked with man's use of fossil fuels and exploitation of the land. . . . If carbon dioxide continues to increase, the study group finds no reason to doubt that climate changes will result and no reason to believe that these changes will be negligible. . . . A wait-and-see policy may mean waiting until it is too late."[53]

Upon issuing the report, the academy noted that the members of Charney's committee "were chosen for their special competencies and with regard for appropriate balance." It also commented: "This report has been reviewed by a group other than the authors according to procedures approved by a Report Review Committee consisting of members of the National Academy of Sciences, the National Academy of Engineering, and the Institute of Medicine."[54] Thus, the Charney committee's conclusion that there is "no reason to believe" that climate changes will be negligible enjoyed substantial scientific support.

Also in 1979, the U.S. Senate Committee on Governmental Affairs held a hearing on "Carbon Dioxide Accumulation in the Atmosphere, Synthetic Fuels and Energy Policy," during which Harvard scientist Roger Revelle stated: "Should we take the CO_2 effect of the various energy strategies into direct account in our decision-making processes? I would answer unequivocally, yes: the question is, how?" At the same hearings, Stephen Schneider, speaking of observed and projected increases in atmospheric CO_2 and the potential impact on the Greenland and West Antarctic ice sheets, testified: "And the concern has come about that should the CO_2 increase cause an increase in global temperature of a [Celsius] degree or two, that might lead to a larger increase in the polar regions, perhaps 5 degrees. And that 5 degrees brings the margins of this West Antarctic ice sheet to the melting point. And the concern is: if these ice sheets broke up then some of the landed ice—ice that is on the [Antarctic] continent—might slip rather quickly into the sea. Perhaps, some have said, this could take only decades. Others say: No. It would take centuries.

There is considerable controversy. This issue is, if you will, the ultimate consequence of the CO_2 question. And it remains shrouded in debate."[55]

Two years later, in 1981, the *New York Times*'s Walter Sullivan wrote about a climate study by NASA's Goddard Institute for Space Studies that was led by James E. Hansen and published in *Science:*

> A team of Federal scientists says it has detected an overall warming in the earth's atmosphere extending back to the year 1880. They regard this as evidence of the validity of the "greenhouse" effect, in which the increasing amounts of carbon dioxide cause steady temperature increases.
>
> The seven atmospheric scientists predict a global warming of "almost unprecedented magnitude" in the next century. It might even be sufficient to melt and dislodge the ice cover of West Antarctica, they say, eventually leading to a worldwide rise of 15 to 20 feet in the sea level. In that case, they say, it would "flood 25 percent of Louisiana and Florida, 10 percent of New Jersey and many other lowlands throughout the world" within a century or less.[56]

In June 1988, Hansen testified to a U.S. Senate committee that he was "99 percent certain" that the climate warming of about 1 degree Fahrenheit over the past one hundred years "was not a natural variation but was caused by a buildup of carbon dioxide and other artificial gases in the atmosphere." Speaking before the committee, chaired by Timothy Wirth, a Democrat from Colorado, Hansen said, "It is time to stop waffling" and acknowledge "that the greenhouse effect is here."[57]

Thus, throughout the 1970s and 1980s, substantial scientific apprehension about the impact of human-induced greenhouse emissions on the Earth's climate had been reported. In response to this concern, in 1988 the United Nations Environment Program and the World Meteorological Organization founded the Intergovernmental Panel on Climate Change "to assess the scientific information that is related to the various components of the climate change issue, such as emissions of major greenhouse gases," and "to enable the environmental and socio-economical consequences of climate change to be evaluated."[58] The IPCC is composed of about 2,500 scientists and other specialists from around the world — all of whom work pro bono — who have expertise in a broad range of climate-related disciplines.

Two years after it was established, the IPCC issued its first assessment report. Though the 1990 report was the most tentative scientifically of the four assessment reports that the IPCC has published to date (see chapter 4), it nevertheless was "certain" that human-induced greenhouse emissions were enhancing the natural greenhouse effect: "We are certain of the following. . . . Emissions resulting from human activities are substantially increasing the atmospheric concentrations of the greenhouse gases: carbon dioxide, methane, chlorofluorocarbons (CFCs) and nitrous oxide. These increases will enhance the greenhouse effect, resulting on average in an additional warming of the Earth's surface."[59]

The 1990 IPCC report also projected increases in global temperatures due to a human-induced increase in atmospheric greenhouse gases. Under the IPCC's "Business as Usual" scenario — that is, with little to no reductions in such gases, roughly what Lomborg prescribes — average temperatures would increase 3–6°C (5.4–10.8°F) by year 2100.[60] Likewise, and contrary to what Lomborg argues in *Cool It* — that large CO_2 cuts "made now will have only a rather small and insignificant impact far into the future"[61] — the IPCC reported in 1990 that progressively increasing levels of controls on greenhouse emissions would lead to correspondingly lower increases in global temperatures. Thus, for the IPCC, if not for Lomborg — according to projections that remained fairly steady over the course of its four assessment reports, from 1990 to 2007 — not reducing or reducing greenhouse emissions and by how much could have a big impact on how much the Earth will warm throughout this century.[62]

A year after the IPCC's 1990 assessment report, the U.S. National Academy of Sciences (NAS) issued an important booklet on global warming titled, *Policy Implications of Greenhouse Warming*. The NAS issued projections of global temperatures relative to increases in greenhouse emissions and, in doing so, issued a warning: "At their present level of development, GCMs [general circulation models] project that an increase in greenhouse gas concentrations equivalent to a doubling of the preindustrial level of atmospheric CO_2 would produce global average temperature increases between 1.9 and 5.2°C (3.4 and 9.4°F). The larger of these temperature increases would mean a climate warmer than any in human history. The consequences of this amount of warming are unknown and could include extremely unpleasant surprises."[63]

Roughly consistent with what the IPCC and NAS had reported by 1991, the 2007 IPCC assessment report projected that global average temperatures would increase by a likely range of 1.1–6.4°C (2–11.5°F) by year 2100.[64] Thus, over the twenty-year period in which the IPCC studied the relationship between man-made greenhouse emissions and a warming Earth, the range of projected temperature increases remained relatively constant, as did the correlation between lower emissions and lower temperatures, and higher emissions and higher temperatures. This indicates that Lomborg's argument—that global warming isn't a big problem and that we needn't bother much with reductions in greenhouse emissions—from the beginning was formulated outside a scientific consensus that projected unprecedented warming with potentially catastrophic consequences if greenhouse emissions were not significantly reduced.

One might have thought that this consensus would have compelled Lomborg to meet a high burden of evidence to support his case that global warming was no catastrophe and that governments needn't prioritize reductions in greenhouse emissions. On the contrary, given the antienvironmentalism of the new American president at the time, the Republican-led right-wing Congress, and the U.S. news media's self-imposed obligation to "balance" political and scientific realities with alternative possibilities—however factually challenged—Lomborg's "skeptical environmentalism" was well positioned to exploit the political and cultural interests that welcomed his book in the United States in fall 2001. Whatever explains Lomborg's motivations and the success of *The Skeptical Environmentalist*, it functions for our purposes as a necessary introduction to his 2007 book, *Cool It: The Skeptical Environmentalist's Guide to Global Warming.*

two
ON POLAR BEARS

Bjørn Lomborg began *Cool It* with a brief chapter on polar bears. He wanted to demonstrate — right off the bat — that what he considered to be an exaggerated threat to polar bears from global warming was a case study of how the threat of global warming itself is exaggerated. "But the real story of the polar bear," Lomborg wrote, "is instructive," since "once you look closely at the supporting data, the narrative [that the bears are threatened] falls apart." Lomborg's exposé of the alleged misuse of polar bear data "encapsulates the broader problem with the climate-change concern," and supports his "skeptical environmentalist's" perspective on major environmental issues.[1] Following Lomborg's cue, we thus begin our own close reading of Lomborg's Theorem (that global warming is no catastrophe) by utilizing Lomborg's chapter on polar bears as our own case study for some early insight into the scholarship of the most influential global warming skeptic of the past decade.

In the polar bear chapter, Lomborg accused *Time* magazine, Al Gore, the World Wildlife Fund, and the 2004 Arctic Climate Impact Assessment (ACIA) of exaggerating the threat to polar bears from global warming. About *Time*, Lomborg cited its noteworthy 2006 cover story — "Global Warming: Be Worried. Be Very Worried"[2] — and wrote: "The heartbreaking image on the cover was of a lone polar bear on a melting ice floe, searching in vain for the next piece of ice to jump to. *Time* told us that due to global warming bears 'are starting to turn up drowned' and that at some point they will become extinct."[3]

About Gore's book, *An Inconvenient Truth*, Lomborg wrote: "Al Gore shows a picture similar to *Time*'s and tells us 'a new scientific study shows that, for the first time, polar bears have been drowning in significant numbers.' " Lomborg then wrote that "over the past few years," such stories have "cropped up many times," and were based "first on a World Wildlife Fund report in 2002 and later on the Arctic

Climate Impact Assessment from 2004." He asserted that both of these reports "relied extensively on research published in 2001 by the Polar Bear Specialist Group of the World Conservation Union."[4]

Having identified a chain of information that begins with the World Conservation Union (formally known as the International Union for the Conservation of Nature [IUCN]) and ending with the accusation that Gore and *Time* exaggerated the plight of polar bears, Lomborg launched his exposé by arguing that, contrary to these reports, the polar bear population has thrived:

> But what this group [IUCN] really told us was that of the twenty distinct subpopulations of polar bears, one or possibly two were declining in Baffin Bay; more than half were known to be stable; and two subpopulations were actually *increasing* around the Beaufort Sea. Moreover, it is reported that the global polar-bear population has *increased* dramatically over the past decades, from about five thousand members in the 1960s to twenty-five thousand today, through stricter hunting regulation. Contrary to what you might expect—and what was not pointed out in any of the recent stories—the two populations in decline come from areas where it has actually been getting colder over the past fifty years, whereas the two increasing populations reside in areas where it is getting warmer. Likewise, Al Gore's comment on drowning bears suggests an ongoing process getting ever worse. Actually, there was a single sighting of four dead bears the day after "an abrupt windstorm" in an area housing one of the increasing bear populations.[5]

Reading this passage for the first time, one might presume that Lomborg was summarizing information from the 2001 report by the IUCN's Polar Bear Specialist Group. But only the first sentence from the passage above is from that report. And Lomborg's claims in that sentence about "stable" and "increasing" polar bear populations only partially reflect what the IUCN report actually said.

The chart in the IUCN report from which Lomborg apparently got this information includes other relevant columns (to which I will refer in a moment) that he ignored. And the IUCN, in fact, qualified its estimates of polar bear populations (called "abundance estimates") on the chart by noting that "abundance estimates are based on the best available data for each population, which ranges from little or no information to detailed inventory studies."[6] And "abundance estimates" are essentially head counts, with few additional implications in this case about "stable" polar bear populations — a word used by Lomborg that does not appear in the IUCN chart. In addition, the headline to the chart noted that the "status" of the polar bear populations — referring to the designation at the top of the only column that Lomborg cited (and which indicated whether bear populations had increased, decreased, or were stationary) — was determined from abundance estimates taken in 1999–2000, and compared to abundance estimates from 1995–96.[7] The five-year span between these estimates is a relatively brief period to be making supposedly definitive claims about "stable" polar bear populations, as Lomborg did.

Furthermore, one of the columns in the chart that Lomborg ignored is designated "Environmental Concerns"; hence, Lomborg neglected to mention that the IUCN chart listed global warming as an environmental concern in eleven of the twenty Arctic subpopulations of polar bears.[8] And among the eleven subpopulations of polar bears that Lomborg described as "stable" ("stationary," as worded in the report), the IUCN identified global warming as having a harmful effect on arctic sea ice or the bears in six of those subpopulations. Also, while the IUCN reported that the Western Hudson Bay subpopulation of polar bears is "stationary," it described the same subpopulation of bears as undergoing a "decline" in the "condition" of the bears because of global warming:

> Over the past two decades, the condition of adult male and female bears and the proportion of independent yearling bears caught during the open water season have declined significantly (Derocher and Stirling 1992, Stirling and Lunn 1997, Stirling et al. 1999). Over the same period of time, the date of break-up of the sea-ice in western Hudson Bay has advanced by two weeks (Stirling et al. 1999), which is probably due to spring air temperatures in the region warming at a rate of 0.2–0.3°C per decade since 1950 (Skinner et al. 1998). Stirling et al. (1999) documented that the timing of [sea-ice] break-up was positively correlated with the condition of adult females (i.e., the earlier the break-up

> the poorer the condition of the bears) and suggested that the declines in the various parameters measured in the polar bears have resulted from the trend toward earlier break-up, which in turn appears to be due to the long-term warming trend in spring temperatures.[9]

Thus, given the early breakup of sea ice as the apparent result of global warming, and the observed poor physical condition of the bears that the IUCN report linked to the early breakup of sea ice, it seems inaccurate to describe the Western Hudson Bay subpopulation as "stable," as Lomborg did, even though the polar bear head counts were roughly the same ("stationary," in the report's words) from 1996 to 2000.

Furthermore, the same IUCN report had this to say—and which Lomborg ignored—about global warming, sea ice, and polar bears throughout the Arctic region: "Anthropogenic [man-made] and natural changes in arctic environments as well as new recognition of the shortcomings of our knowledge are increasing the uncertainties of polar bear management. Higher temperatures and erratic weather fluctuations, apparent symptoms of global climate change, are increasing across the range of polar bears. Following the predictions of climate modelers, such changes have been most prevalent in Arctic regions (Stirling and Derocher 1999, Stirling and Lunn 1999), and already have altered local and global sea-ice conditions (Gloersen and Campbell 1991, Vinnikov et al. 1999). Because changes in sea-ice are known to alter polar bear numbers and productivity (Stirling and Lunn 1997, Stirling et al. 1999), effects of global climate changes can only increase future uncertainty and may increase risks to the welfare of polar bear populations."[10] There is thus little if any basis in this passage—or in the IUCN report overall—upon which to characterize polar bear populations in the Arctic as "stable," given the emphasis in the IUCN report on the "risks" that polar bears are encountering from global warming, and the deteriorating condition of the bears that already had been reported.

Similar problems arise upon reviewing Lomborg's use of the 2004 ACIA. The ACIA was a large research project in which three hundred scientists participated from eight Arctic countries—the United States,

Canada, Russia, Sweden, Norway, Finland, Denmark, and Iceland. Lomborg dismissed the ACIA's findings with respect to polar bears because, as he wrote, it "relied extensively on research published in 2001 by the Polar Bear Specialist Group of the World Conservation Union"[11] — that is, the 2001 IUCN report referenced above. Except for providing the URL of the IUCN's Polar Bear Specialist Group,[12] Lomborg presented no additional details to support this claim. And in the three most detailed sections on polar bears in the thousand-page ACIA report,[13] the IUCN is cited only once, and that was for a 1998 report, not the 2001 report. Thus, in addition to what Lomborg ignored from the 2001 IUCN report, he also summarily dismissed important research findings and summaries (excerpted below) from the 2004 ACIA report pertaining to polar bears and global warming.

In a section titled "Polar Bears" in a chapter on "Marine Systems," the ACIA briefly described the singular importance of Arctic sea ice as polar bear habitat (references are included below as author-date citations, as listed in the original text):

> The polar bear, the pinnacle predator, has a circumpolar navigation and is dependent on sea ice to provide for most of its needs (Ferguson et al., 2000a, b; Mauritzen M. et al., 2001; Stirling et al., 1993). Polar bears feed almost exclusively on ice-associated seals (e.g., Lønø, 1970; Stirling and Archibald, 1977; Smith T., 1980). Adult bears can swim quite long distances if required, but mothers with cubs depend on ice corridors to move young cubs from terrestrial denning areas to prime hunting areas on the sea ice (Larsen T., 1985, 1986). Pregnant females dig snow dens in the early winter and give birth several months later. This requires a significant depth of snow, thus females return year after year to land sites that accumulate sufficient snow early in the season. A mother that emerges from the den with her young has not eaten for five to seven months (Ramsay and Stirling, 1988). Therefore, successful spring hunting is essential for the family's survival and largely dictates condition, reproductive success, and survival for all polar bears (e.g., Stirling and Archibald, 1977). Factors that influence the distribution, movement, duration, and structure of sea ice profoundly affect the population ecology of polar bears, not least due to their influence on the principal prey species, ringed seal (*Phoca hispida*) (Stirling and

> Øritsland, 1995; Stirling et al., 1999). The global polar bear population is estimated at 22,000 to 27,000 (IUCN, 1998).[14]

This passage was supported by about a dozen peer-reviewed scientific studies; only the last sentence was referenced to the IUCN (that is, to a 1998 report, not the 2001 report). In a related passage, the 2004 ACIA report described the effects of global warming on sea ice and polar bears:

> Changes in the extent and type of sea ice will affect the distribution and foraging success of polar bears. The earliest impact of warming had been considered most likely to occur at the southern limits of their distribution, such as James and Hudson Bays (Stirling and Derocher, 1993), and this has now been documented (Stirling et al., 1999). Late sea-ice formation and early breakup means a longer period of annual fasting for polar bears. Reproductive success is strongly linked to their fat stores; females in poor condition have smaller litters and smaller cubs, which are less likely to survive, than females in good condition. There are also concerns that direct mortality rates are likely to increase with the climate-change scenarios projected by the ACIA-designated models. For example, increased frequency or intensity of spring rain could cause dens to collapse resulting in the death of the female as well as the cubs. Earlier spring breakup of ice could separate traditional den sites from spring feeding areas, and young cubs forced to swim long distances from breeding areas to feeding areas would probably have a lower survival rate. *It is difficult to envisage the survival of polar bears as a species given a zero summer sea-ice scenario.* Their only option would be a terrestrial summer lifestyle similar to that of brown bears, from which they evolved.[15] [Emphasis added]

This passage likewise did not cite the 2001 IUCN report; so much for Lomborg's claim that the 2004 ACIA's findings on polar bears "relied extensively" on the 2001 IUCN report. And Lomborg ignored the ACIA's prognosis that "it is difficult to envisage the survival of polar bears as a species" pursuant to a loss of Arctic sea ice in summer. Even though Lomborg did allude to this paragraph, he misrepresented what it said. In his chapter on polar bears in *Cool It*, Lomborg wrote: "Yes, it is likely that disappearing ice will make it harder for polar bears to continue their traditional foraging patterns and that they will increasingly take up a lifestyle similar to that of brown bears, from which they evolved. They may eventually decline, though dramatic declines seem unlikely."[16] Thus, while the ACIA section in question reported that, due to disappearing sea ice, "it is difficult to envisage the survival of polar bears as a species," Lomborg reported, while referring to and sourcing the very same section in the ACIA document, that "dramatic declines [in the polar bear population] seem unlikely."[17]

While disparaging or dismissing major scientific sources on polar bears, much of Lomborg's strangely upbeat assessment is referenced to offbeat sources. For example, as I've already noted, Lomborg wrote: "Contrary to what you might expect — and what was not pointed out in any of the recent stories — the two [polar bear] populations in decline come from areas where it has actually been getting colder over the past fifty years, whereas the two increasing populations reside in areas where it is getting warmer."[18] Thus, "contrary to what you might expect," as he put it, Lomborg argued that the polar bears were thriving in warmer weather. What is Lomborg's source to support this claim? For one thing, Lomborg's assertion is inconsistent with the 2001 IUCN report, which observed that "the condition of adult male and female [polar] bears" in the southerly Western Hudson Bay subpopulation "have declined significantly" where "spring air temperatures" have "warm[ed] at a rate of 0.2–0.3°C per decade since 1950." The IUCN also reported that these "declines in the various parameters measured in the [Western Hudson Bay] polar bears have resulted from the trend toward earlier break-up [of sea ice], which in turn appears to be due to the long-term warming trend in spring temperatures."[19] And this information is consistent with the 2004 ACIA report, which noted that "the condition of adult male and female polar bears has declined in Hudson Bay since the early 1980s, as have birth rates and the proportion of first-year cubs in the population," and "that the proximate cause of these changes in physical and reproductive parameters is a trend toward earlier break-up of the sea ice [due to warming], which has resulted in the bears coming ashore in poorer condition."[20]

Lomborg, however, ignored these sources and these findings, and argued differently by citing two other sources. The first one is a commentary that was posted by Patrick J. Michaels of the libertarian think tank Cato Institute at the group's Web site (cato.org) in November 2004. Titled "Polar Disasters: More Predictable Distortions of Science," the post began: "November has been quite a month for climate disaster stories! First, *Nature* magazine reports that the Antarctic food chain is all out of whack, with krill populations crashing around the South Orkney Islands because of global warming. Then a new federally funded [*sic*] 'Arctic Climate Impact Assessment' (ACIA) comes along, predicting the upcoming extinction of polar bears and the death of Inuit culture."[21] Note the willy-nilly accusation that *Nature* and the ACIA are peddling distorted science. Note also that even Michaels observed that the 2004 ACIA report projected the possible extinction of polar bears due to global warming — a feature of that report that Lomborg ignored.

Near the end of his commentary, Michaels argued that the ACIA left out the following alleged facts about the decline of the polar bear: "Where the polar bear populations are in decline — around Baffin Bay (the region between Canada and Greenland), temperatures are also going down, big time. And the area where temperatures are rising the most — in the Pacific region bordering on Alaska and Siberia, polar bear populations are increasing."[22] Michaels supported these comments with no references, and he offered no further analysis or explanation, other than noting that he observed this himself, with an inference that this personal observation undermines the polar bear and global warming dogmatists.

The 2001 IUCN report, however, provided the likely explanation for Michaels's observation about declining temperatures and declining polar bear populations in Baffin Bay: "This [polar bear] population [in Baffin Bay] is shared with Greenland, which does not limit the number of polar bears harvested. Based on the preliminary population estimate and the most recent harvest [hunting] information

(Born 2000), it appears the [polar bear] population may be over-harvested. Better information on population numbers and validation of the Greenland harvest data are required to clarify the status of this population."[23] Thus, it would seem that the polar bear population in Baffin Bay is declining due to unregulated hunting, and not to any correlation with colder temperatures that might undermine the prevailing scientific assessments of the harmful impact of global warming on polar bears.

Lomborg's second source was a report published in 2000 in the *International Journal of Climatology* and authored by Rajmund Przybylak.[24] Lomborg cited this analysis, in addition to the commentary by Michaels, to support his claim that polar bear populations are decreasing where temperatures have been trending downward, and increasing where temperature trends are up. However, the Przybylak paper never mentions polar bears, and thus provides no additional evidence, beyond what Lomborg himself wrote in *Cool It*, to support his claim that polar bears appear to be thriving in Arctic regions that have grown warmer.[25]

Undeterred, Lomborg provided another erroneous claim that appeared to undermine the prevailing scientific view: "The best-studied polar bear population lives on the western coast of Hudson Bay. That its population has declined 17 percent, from 1,200 in 1987 to under 950 in 2004, has gotten much press. Not mentioned, though, is that since 1981 the population had soared from just 500, thus eradicating any claim of a decline."[26]

Lomborg supported the last sentence of this statement with three sources, the first being a study led by Ian Stirling, a leading polar bear expert, in the science journal *Arctic* (Stirling, Lunn, and Iacozza, 1999). The ninth page of the study includes a chart which indicates that the polar bear population in the Western Hudson Bay increased from about 500 in 1981 to 1,200 in 1997.[27] Lomborg argued that such figures "eradicate any claim of a decline" in polar bears in the Western Hudson Bay during this period. However, the first sentence of the first page of Stirling's report states: "From 1981 through 1998, the condition of adult male and female polar bears has declined significantly in western Hudson Bay, as have natality [birth rates] and the proportion of yearling cubs caught during the open water period that were independent at the time of capture." The researchers attributed the "proximate cause" of the decline to "earlier break-up" of sea ice "correlated with rising spring air temperatures" due to "a long-term warming trend in April–June atmospheric temperatures."[28]

Also, in his reference citing the Stirling study Lomborg cited "Amstrup et al., 2006: slide 44" as allegedly "confirm[ing]" Lomborg's interpretation of the Stirling study on the polar bear population in the Western Hudson Bay. However, while slide 44 in the Amstrup paper shows an increase in the polar bear population in Western Hudson Bay from the mid-1980s to the late-1980s, it also reports: "The WHB [Western Hudson Bay] polar bear population has declined from over 1100 individuals in 1988 to fewer than 950 individuals in 2004."[29] Furthermore, Amstrup almost certainly derived these figures from a 2005 report by the IUCN's Polar Bear Specialist Group, which reported:

> 3. Status of the Western Hudson Bay (WH) population analysis
> The IUCN Polar Bear Specialist Group
> *Recognising* that the largest and best developed scientific database for any polar bear population is the WH database, and
> *Recognising* that the current WH mark-recapture population analysis has used multiple standardized methodologies which produced equivalent estimates, and
> *Recognising* that the analysis results are consistent with independent population simulation results, and
> *Recognising* that the data used for these estimates have been carefully checked and validated, and
> *Noting that the decline of WH polar bears from approximately 1200 in 1987 to less than 950 in 2004 is conclusive,*
> *And accepting that the decline was due to a combination of anthropogenic removals (defence and harvest kills) and reduced demographic rates from climate warming,* therefore
> *Recommends* that appropriate management action be taken without delay.[30] [Emphasis added]

While this 2005 report from the IUCN's Polar Bear Specialist Group, the leading scientific authority on polar bears, would appear to be conclusive with respect to a decline in the polar bear population in the two decades prior to the publication of Lomborg's 2007 book, *Cool It*, Lomborg viewed the brief increase in the polar bear population in the mid-1980s as the definitive statistic, both noted and disregarded the consensus scientific view that the polar bear population had declined in the two-decade period that followed the mid-1980s, and disregarded altogether the IUCN's assessment that this most recent decline was due to global warming.

Furthermore, the 2001 report of the IUCN's Polar Bear Specialist Group reported: "Over the past 20–25 years, a decline in the reproductive rate and cub survival of polar bears in Western Hudson Bay has been documented."[31] Though Lomborg cited this report, he ignored this finding. Finally, a November 2007 study in the *Journal of Wildlife Management*, published shortly after *Cool It*'s publication in September, reported: "The Western Hudson Bay polar bear population declined from 1,194 in 1987 to 935 in 2004."[32]

Thus, Lomborg's claim that the polar bear population "had soared" since 1981 was highly selective and misleading, since he neglected to give more weight to, or adequately divulge the contents of, major scientific reports of the more recent decline in the Western Hudson Bay subpopulation of polar bears, which the reports attributed to global warming. Lomborg's approach is thus comparable to a doctor telling a patient that, despite a twenty-year serious chronic illness, the patient is not actually sick on the grounds that the patient's health was better twenty-five years ago.

While disregarding authoritative reports of a decline in the condition and numbers of polar bears due to global warming, Lomborg argued that even if we accepted "the story" of a polar bear decline "at face value," the reason for that decline is a lack of restrictions on hunting polar bears and not global warming. Thus, after writing that "since 1981 the [polar bear] population had soared from just 500, thus eradicating any claim of a decline," Lomborg wrote: "Moreover, nowhere in the news coverage is it mentioned that 300 to 500 [polar] bears are shot each year, with 49 shot on average on the west coast of

Hudson Bay. Even if we take the story of [polar bear] decline at face value, it means we have lost about 15 bears to global warming each year, whereas we have lost 49 each year to hunting."[33] Thus, for Lomborg, polar bear hunting threatens polar bears more than greenhouse emissions and global warming. He clearly argues this point in *Cool It:*

> We are being told that the plight of the polar bears shows "the need for stricter curbs on greenhouse-gas emissions linked to global warming." Even if we accept the flawed idea of using the 1987 population of polar bears around Hudson Bay as a baseline, so that we lose 15 bears each year, what can we do? If we try helping them by cutting greenhouse gases, we can at the very best avoid 15 bears dying. We will later see that realistically we can do not even close to that much good—probably we can save about 0.06 bears per year. But 49 bears from the same population are getting shot each year, and this we can easily do something about. Thus, if we really want a stable population of polar bears, dealing first with the 49 shot ones might be both a smarter and a more viable strategy. Yet it is not the one we end up hearing about.[34]

Lomborg's only source for the number of polar bears killed by hunting is the same chart referred to earlier from the 2001 IUCN report. However, according to the chart, the overall number of polar bears killed annually (designated "Mean Annual Kill") is less than the allowable sustainable number of polar bears killed ("Sustainable Kill") in seven of the ten geographical regions where such statistics are reported. This includes the Western Hudson Bay region, where the "Mean Annual Kill" is forty-nine polar bears (as Lomborg reported) but where the "Sustainable Kill" is fifty-two polar bears per year. Since the same chart also reports that the "total estimate of world abundance" (head counts) for polar bears is 21,500 to 25,000, it seems plausible that a mean annual kill of 300 to 500 polar bears due to hunting would be "sustainable," and thus by itself would not necessarily be a serious threat to the polar bear species.[35] More importantly, Lomborg's analysis disregards the fundamental issue of disappearing Arctic sea ice, and thus the essential habitat of the polar bear, which is threatened by greenhouse emissions and global warming.

Tighter hunting regulations, though possibly desirable, obviously would have no effect on preserving Arctic sea ice, whereas reducing greenhouse emissions ultimately might.

Lomborg also supported his claim that polar bears are threatened more by hunting than by global warming by citing the comments of a Canadian scientist. Lomborg wrote: "In 2006, a polar-bear biologist from the Canadian government summed up the discrepancy between the [hunting] data and the PR [that polar bears are threatened by global warming]." Lomborg then quoted the scientist: "It is just silly to predict the demise of the polar bear in 25 years based on media-assisted hysteria." The polar bear biologist also noted that the bears "are not going extinct, or even appear to be affected at present."[36] However, Lomborg's source for these statements was a four-hundred-word opinion piece published in May 2006 in the *Toronto Star* by Dr. Mitchell Taylor, identified as a polar bear biologist for the government of Nunavut (Canada).[37] While Taylor may be a fine scientist, it is not compelling scholarship on Lomborg's part to contradict the extensive data and analysis in the thick scientific reports issued by the IUCN and the Arctic Climate Impact Assessment—which include well-documented threats to polar bears from human-induced global warming and the disappearance of Arctic sea ice—with a brief opinion piece from a newspaper that is not peer-reviewed.

Also, since Lomborg apparently was inclined to argue his case in part by citing newspaper items, there were a number of relevant news articles published prior to May 2006 that he could have cited, in addition to the brief commentary from the *Toronto Star*. For example, Toronto's *Globe and Mail* reported in November 2002 that the "vast expanse of permanent ice that has characterized the Arctic Ocean for millennia is fated to disappear far faster than anyone imagined, and will certainly be gone before the century is out, says a NASA satellite study." The "startling" NASA survey, conducted from 1978 to 2000, "shows that an area of ice roughly as large as Alberta is vanishing every decade as the climate warms." The rate of melting of the Arctic ice, NASA found, was "roughly 9 per cent a decade" and "speeding up." The lead author of the study, Josefino Comiso, a senior scientist at NASA's Goddard Space Flight Center in Maryland, said, "This year we had the least amount of permanent ice cover ever observed." The *Globe and Mail* article continued: "The findings have huge implications for global climate patterns. Arctic snow and ice play a key role in controlling the planet's temperature. They act as insulation, keeping heat and moisture in the land and ocean and out of the atmosphere. But once the ice and snow are gone, that dynamic will end and this will affect climate all over the planet in ways scientists have not yet begun to fathom. The Arctic itself, so long forsaken, is likely to become humid and warm. Animals and fish that thrive on the permanent ice and snow—polar bears, for example—are likely to die off, unable to survive the heat."[38] The NASA study apparently is the kind of "PR" about global warming that Lomborg complains about, and may explain why—of the two newspaper items from Toronto—he ignored this one.

In December 2002, Kenneth Chang of the *New York Times*, reporting from a conference of the American Geophysical Union, summarized the findings of a paper delivered by scientists from the National Snow and Ice Data Center in Boulder, Colorado: "The melting of Greenland glaciers and Arctic Ocean sea ice this past summer reached levels not seen in decades."[39] In September 2003, the *Independent* reported that "the largest ice shelf in the Arctic"—the Ward Hunt ice shelf, on the north coast of Ellesmere Island in Canada—"a solid feature for at least 3,000 years, has broken in two and climate change is to blame, say American and Canadian scientists."[40] A month later, in October 2003, Agence France Presse, citing a study by NASA's Goddard Institute for Space Studies in New York, reported that "the polar ice cap is melting at an alarming rate due to global warming," with satellite images showing that the ice cap "has been shrinking by 10 percent per decade over the past quarter century," and "the extent of Arctic sea ice that remains frozen all year reached record lows in 2002 and 2003."[41] Lomborg ignored these news reports, and the scientific studies that prompted them, which are clearly relevant to the impact of global warming on the Arctic sea-ice habitat of polar bears.

The following news article, however, is one that Lomborg did cite, and it is worth summarizing in some detail. In November 2004, reporter Juliet Eilperin of the *Washington Post* began her article on the just-released Arctic Climate Impact Assessment:

> Global warming could cause polar bears to go extinct by the end of the century by eroding the sea ice that sustains them, according to the most comprehensive international assessment ever done of Arctic climate change. The thinning of sea ice—which is projected to shrink by at least half by the end of the century and could disappear altogether, according to some computer models—could determine the fate of many other key Arctic species, said the Arctic Climate Impact Assessment, the product of four years of work by more than 300 scientists.
>
> Bears are dependent on sea ice because they use it to hunt for seals, which periodically pop up through breathing holes in the ice. Because the ice has broken up earlier and earlier in the year over the past few decades, polar bears are deprived of crucial hunting opportunities. The uncertain fate of the world's largest non-aquatic carnivores—as well as the future of other animals and humans who live in the Arctic—was sketched in stark relief yesterday by the 139-page [ACIA summary] document.
>
> The report offered a broad picture of the evidence that climate change has disproportionately affected far northern latitudes. The researchers concluded that some areas in the Arctic have warmed 10 times as fast as the world as a whole, which has warmed an average of 1 degree Fahrenheit over the past century.

When citing this report from the *Washington Post* in *Cool It*, Lomborg mentioned none of these details. The *Post* article continued:

> The sea ice in Hudson Bay, Canada, now breaks up 2½ weeks earlier than it did 30 years ago, said Canadian Wildlife Service research scientist Ian Stirling, and as a result female polar bears there weigh 55 pounds less than they did then. Assuming the current rate of ice shrinkage and accompanying weight loss in the Hudson Bay region, bears there could become so thin by 2012 they may no longer be able to reproduce, said Lara Hansen, chief scientist for the World Wildlife Fund.

Nor did Lomborg refer to any of these details while citing the *Washington Post* article, which continued:

> Arctic residents have already detected changes in polar bears' behavior. Jose Kusugak, president of the Canadian Inuit political association, said at a news conference that within the past two years he witnessed a polar bear "stock up on caribou" because it was deprived of seals. Hudson Bay residents now complain the bears are coming onto land more often, forced to seek sustenance in a habitat where they are less well adapted. Polar bears are not the only Arctic animals in trouble. The ringed seals that bears eat, and that humans hunt, are also dependent on the sea ice to rest, give birth, nurse and feed. "You have organisms that have been pushed beyond their limits," said James McCarthy, director of the Harvard University Museum of Comparative Zoology.[42]

Lomborg ignored these details as well, which prompts the question: What did Lomborg say about this report?

Toward the end of the *Post* story, and after all of the details above had been given, it noted that "environmentalists said [the Arctic Climate Impact Assessment] shows the need for stricter curbs on greenhouse gas emissions linked to global warming." Ignoring everything else in Eilperin's article, Lomborg referred only to this part of it to complain that "we are being told that the plight of the polar bear shows 'the need for stricter curbs on greenhouse gas emissions linked to global warming.' "[43] While the claim is narrowly accurate, it is also misleading, since Lomborg is arguing that "we are being told" to reduce greenhouse emissions for the sake of an exaggerated threat to polar bears, when in fact he withheld evidence from the same *Washington Post* article indicating that the bears are indeed threatened. And Lomborg withheld this evidence after writing one page earlier in *Cool It* that "we hear vastly exaggerated and emotional claims" about polar bears.[44]

Lomborg also ignored a report, published in the *Guardian* in September 2005, concerning the deterioration of Arctic sea ice—which is what the 2004 ACIA cited as threatening the polar bears with extinction: "Global warming in the Arctic could be soaring out of control, scientists warned yesterday, as new figures revealed that melting of sea ice in the region has accelerated to record levels. Experts at the US National Snow and Data Centre in Colorado fear the region is locked into a destructive cycle with warmer air melting more ice, which in turn warms the air further. Satellite pictures show the extent of Arctic sea ice this month dipped some 20% below the long term average for September—melting an extra 500,000 square miles, or an area twice the size of Texas. If current trends continue, the summertime Arctic Ocean will be completely ice-free well before the end of this century." The article noted that "the decline [in Arctic sea ice] threatens wildlife in the region, including polar bears that spend the summer on land before returning to the ice when it reforms in winter."[45]

A year later, in September 2006, in an article titled "Even in Winter, Arctic Ice Melting," the *San Francisco Chronicle* reported: "The vast expanses of ice floating in the Arctic Sea are melting in winter as well as in the summer, likely because of global warming, NASA scientists said Wednesday." The article noted that "particularly hard hit would be the polar bears, which live on the ice."[46]

In *Cool It*'s opening chapter on polar bears, Lomborg also cited what he viewed as exaggerated reports about drowning polar bears to illustrate his argument that the threat from global warming is overstated. In his 2006 book, *An Inconvenient Truth*, Al Gore included a photograph of a mother polar bear and her cub on an ice floe. In the accompanying text, Gore wrote: "The melting of the ice represents bad news for creatures like polar bears. A new scientific study shows that, for the first time, polar bears have been drowning in significant numbers. Such deaths have been rare in the past. But now, these bears find they have to swim much longer distances from floe to floe. In some places, the edge of the ice is 30 to 40 miles from the shore."[47] Lomborg took issue with this statement: "Al Gore's comment on drowning bears suggests an ongoing process getting ever worse. Actually, there was a single sighting of four dead bears the day after 'an abrupt windstorm' in an area housing one of the increasing bear populations."[48]

To support his argument, Lomborg cited a 2005 report issued by the state of Alaska titled, "Potential Effects of Diminished Sea Ice on Open-water Swimming, Mortality, and Distribution of Polar Bears During Fall in the Alaskan Beaufort Sea." This report stated (the words in parentheses are in the original text): "Following an abrupt windstorm, 4 dead bears were seen floating far offshore (versus 0 in all previous years). Those bears are believed to have drowned as a result of the storm." The second of these two sentences would seem to support what Lomborg had reported. However, these sentences were located in a larger paragraph, which, in its entirety, reads (words in parentheses in original): "During September 2004 an unusual number of bears were seen swimming offshore (10 of 51 (20%) versus 12 of 315 (4%) in 1986–2003). Following an abrupt windstorm, 4 dead bears were seen floating far offshore (versus 0 in all previous years). Those bears are believed to have drowned as a result of the storm. The survey has about 10% coverage so it is likely that many other bears also drowned but were not seen."[49]

When Lomborg asserted that only four polar bears had drowned due to an isolated incident, he neglected to mention that, according to his own source, almost as many polar bears (ten) were seen swimming offshore in one month (September 2004) than in the combined previous eighteen years (when twelve polar bears were seen). Further, he omitted the researchers' conclusion that, in September 2004 alone, "it is likely that many other bears also drowned but were not seen." Lomborg's representation of this study is also questionable given that he also disputed Gore's claim that the drowning polar bears suggests an upward trend of drownings, though the same study suggested that polar bear drownings may increase due to melting Arctic sea ice: "We suggest that drowning-related deaths of polar bears may increase in the future if the observed trend of regression of pack ice and/or longer open water periods continues."[50]

This section, and the text and charts of the report, present evidence contrary to Lomborg's claims that polar bear drownings are not a concern and that Gore exaggerated the drowning threat to polar bears. In fact, Gore's account of polar bear drownings was comparable to an article on the same study in the *Sunday Times* (London), ignored by Lomborg, which similarly reported that "scientists for the first time found evidence that polar bears are drowning because climate change is melting the Arctic ice shelf," and "the scientists believe such drownings are becoming widespread across the Arctic, an inevitable consequence of the doubling in the past 20 years of the proportion of polar bears having to swim in open seas."[51] In this case, as in others, Lomborg found a needle of technical accuracy (the study noted a single incident in which four polar bears were drowned) in a haystack of countervailing evidence that Lomborg ignored (drownings and the drowning threat to polar bears have almost certainly increased due to shrinking Arctic sea ice) so as to issue a groundless accusation that Gore had exaggerated the increased risk of climate-related drownings to polar bears.

Having completed his "exposé" of the polar bear data, Lomborg observed that "the polar bear story teaches us three things. First," he wrote, "we hear vastly exaggerated and emotional claims that are simply not supported by the data"; however, the only exaggerations (and omissions) appear to be Lomborg's.[52] His statement is also strange as applied to the 2001 IUCN report and the 2004 Arctic Climate Impact Assessment, which are unexaggerated and unemotional to a fault.

"Second," Lomborg continues, "polar bears are not the only story."[53] Reasoning optimistically, like his mentor, Julian Simon, that the extinction of polar bears (should that occur) would lead to an Arctic glass that is half full, he wrote: "While we hear only about the troubled species, it is also a fact that many species will do *better* with climate change. In general, the Arctic Climate Impact Assessment projects that the Arctic will experience *increasing* species richness and higher ecosystem productivity. It will have less polar desert and more forest. The assessment actually finds that higher temperatures mean more nesting birds and more butterflies. This doesn't make up for the polar bears, but we need to hear both parts of the story" (emphasis in original).[54] Yes, the 2004 ACIA report noted, while assessing long-term trends, that "while there will be some losses in many Arctic areas, movement of species into the Arctic is likely to cause the overall number of species and their productivity to increase, thus overall biodiversity measured as species richness is likely to increase along with major changes at the ecosystem level." It also reported that "warming is very likely to lead to slow northward displacement of tundra by forests, while tundra will in turn displace high-arctic polar desert."[55]

But the ACIA report also noted other "projected impacts on terrestrial ecosystems" and "projected impacts on freshwater ecosystems" that Lomborg ignored. For example, Lomborg lauds the replacement of tundra with forest; however, a chart on the same page that Lomborg cited from the 2004 ACIA about tundra being replaced by forest (p. 998) also reported, under the heading "Albedo Feedback," "The positive feedback of albedo change (due to forest expansion) on climate is likely to dominate over the negative (cooling) feedback from an increase in carbon storage. The albedo reduction due to reduced terrestrial snow cover will be a major additional feedback." This means that the forest expansion in the Arctic that Lomborg characterizes as a desirable impact would almost certainly contribute to an acceleration of global warming.

Likewise, the very next page of the 2004 ACIA (p. 999) stated: "Some of the [ACIA-designated] models project an entirely ice-free Arctic Ocean in summer by the end of the 21st century. Greater expanses of open water will also increase the positive feedback of albedo change to climate." As in Lomborg's simplistic bean counting of polar bear populations, whereupon he overlooks the reported facts of underweight bears progressively losing their Arctic sea-ice habitat, Lomborg simplistically favors the idea of an Arctic forest replacing Arctic tundra, while overlooking the likelihood that this and other impacts of global warming (including the melting of Arctic ice) will greatly accelerate global warming.

Furthermore, the 2004 ACIA also reported on these very same pages (pp. 998–99) that the survival of several indigenous Arctic species, in addition to polar bears, would be threatened by the arrival of invasive species from the south—another warming-related impact that Lomborg spins as an offsetting net positive impact, even as the polar bears increasingly are at risk. Here is what the ACIA reported (parentheses in original): "Specialist species adapted to the cold arctic climate, ranging from mosses, lichens, vascular plants, some herbivores (lemmings and voles) and their predators, to ungulates (caribou and reindeer), are at risk of marked population decline or extirpation locally. This will be largely as a consequence of their inability to compete with species invading from the south." And: "Reduced sea-ice extent and more open water are very likely to change the distribution of marine mammals (particularly polar bears, walrus, ice-inhabiting seals, and narwhals) and some seabirds (particularly ivory gulls), reducing their populations to vulnerable low levels. It is likely that more open water will be favorable for some whales species and that the distribution range of these species is very likely to spread northward."[56]

And while Lomborg noted that "higher [Arctic] temperatures mean more nesting birds and more butterflies,"[57] he neglected to mention that he acquired this information from a set of three charts consecutively arranged, with the middle chart indicating that there will be an increase in the number of ground beetle species.[58] Why mention the birds and butterflies but not the beetles? And why fail to mention, while claiming that "many species will do better with climate change," that the 2004 ACIA reported that "higher Arctic temperatures" mean that, in addition to threatening polar bears, other threatened Arctic species will include caribou, reindeer, walrus, seals, narwhals, lemmings, voles, mosses, and lichens?

While Lomborg sought to demonstrate that the threat to polar bears from global warming is exaggerated by Al Gore and other environmentalists, he also argued that worrying needlessly about an "exaggerated" threat to polar bears "makes us focus on the wrong solutions," which Lomborg described as "stricter curbs on greenhouse-gas emissions linked to global warming" and "large and very expensive CO_2 cuts." Because "many other issues are much more important than global warming," he argues, we "need to remind ourselves that our ultimate goal is not to reduce greenhouse gases or global warming per se but to improve the quality of life and the environment." Thus, if we want "to leave the planet in decent shape for our kids," then "radically reducing greenhouse-gas emissions is not necessarily the best way to achieve" that goal.[59] While Lomborg conceded in his chapter on polar bears that "global warming is real and man-made" and "will have a serious impact on humans and the environment toward the end of this century," in the next sentence he reiterated his dominant theme that "statements about the strong, ominous, and immediate consequences of global warming are often wildly exaggerated, and this is unlikely to result in good policy."[60]

Indeed, Lomborg's first chapter in *Cool It*—"Polar Bears: Today's Canaries in the Coal Mine?"—was valuable as a case study on how the impact of global warming is exaggerated, and how, as a result, bad policy recommendations are made. However, the exaggerations and bad policy recommendations were clearly Lomborg's.

three
ON LOMBORG'S ENDNOTES

Lomborg's earlier and more celebrated book, *The Skeptical Environmentalist*, began just as inauspiciously as *Cool It.* The subtitle to the book is "Measuring the Real State of the World," and the first sentence asks, "What kind of state is the world really in?" By the fourth paragraph, however, Lomborg appears to take the subtitle back: "Of course, it is not possible to write a book (or even lots and lots of books for that matter) which measures the entire state of the world."[1] Why, then, would Lomborg subtitle the book "Measuring the *Real* State of the World"? Is there any difference between the *real* state of the world and the *entire* state of the world? Here is the larger context in which Lomborg describes his book's thesis and how he distinguishes between the *real* world and the *entire* world:

> This book is the work of a skeptical environmentalist. Environmentalist, because I—like most others—care for our Earth and care for the future health and well-being of its succeeding generations. Skeptical, because I care enough to want us not just to act on the myths of both optimists and pessimists. Instead, we need to use the best available information to join others in the common goal of making a better tomorrow.
>
> Thus, this book attempts to measure the real state of the world. Of course, it is not possible to write a book (or even lots and lots of books for that matter) which measures the entire state of the world. Nor is this my intention. Instead, I wish to gauge the most important characteristics of our state of the world—the *fundamentals*. And these should be assessed not on the myths but on the best available facts. Hence, the *real* state of the world.[2] [Emphasis in original]

For Lomborg, then, measuring the *real* state of the world is possible, while measuring the *entire* state of the world is not, even in lots of books. Though it seems that these vague distinctions are essentially useless, as goals they seem, in any event, unattainable by a single non-scientist working outside a peer-review process. Lomborg overcomes these obstacles by issuing what amount to environmental declarations of opinion: "Global warming, though its size and future projections are rather unrealistically pessimistic, is almost certainly taking place, but the typical cure of early and radical fossil fuel cutbacks is way worse than the original affliction, and moreover its total impact will not pose a devastating problem for our future."[3]

Although Cambridge University Press somehow overlooked the questionable nature of Lomborg's scholarly vision and found some merit in these first pages of the manuscript that would ultimately make Lomborg the world's most influential opponent of reducing greenhouse emissions, one would assume that its editors would have scrutinized Lomborg's scholarship—his asserted scientific and environmental facts and the underlying documented sources—before unleashing the Lomborg phenomenon. A simple test in this regard would have been to review at least some of the citations to see whether factual assertions were supported by the sources provided, and thereby assess the merits of Lomborg's foundational claim that environmentalists harbor gloomy "myths" while he alone provides the antidotal "facts" about the *real* state of the world's environment.

If you were an editor at Cambridge University Press, and somehow found Lomborg's unrealistic premise acceptable—that, by himself, he had measured the real state of the Earth's natural environment and found it to be in good shape, contrary to what the major environmental organizations and scientists have found—how would you go about checking nearly three thousand endnotes in order to evaluate the underlying scholarship? One approach would be to start at the beginning.

The first chapter of *The Skeptical Environmentalist*, "Things Are Getting Better," contains about 250 citations. To check a factual assertion for any of these citations, the reader must traverse three sections of the book: the text, which contains the sentences and assertions that are referenced in the endnotes; the "Notes," which contain just the author/date citations, including, on occasion, page numbers; and the "Bibliography," which contains the more fully referenced source, including the full title of the cited periodical or book. Given the essence of Lomborg's scientific dissent—that he uniquely holds and presents the facts—one might have thought that he would have sought to make the validation of his facts as simple and transparent as possible. Instead, Lomborg presented his documentation system in *The Skeptical Environmentalist* as the scholarly equivalent of an obstacle course, seemingly designed to test the limits of an inquisitive reader's perseverance and sanity. The documentation system in *Cool It* is even more challenging, as Lomborg eliminated numbered citations in the text, thus challenging the reader to muster an additional level of resolve by having to identify which sentences or assertions in the text were sourced in the first place. In neither book was Lomborg's documentation system designed to economize the number of pages in the published product (thus reducing the environmental impact), since Lomborg's system significantly added to the number of pages in each book. Although in principle these documentation systems are generally acceptable (though better matched to books with fewer endnotes), in practice they were ill-suited to Lomborg's stated purpose of overthrowing the scientific orthodoxy about the real state of the world's environment—a mission that should have been viewed as bearing the burden of easily and convincingly documenting his rendition of facts.

Also problematic is the degree to which Lomborg's notes inflate the note count in *The Skeptical Environmentalist* without adding to any appreciable scholarship. The first note (in the Preface) is to a quote by former British Prime Minister Benjamin Disraeli ("There are three kinds of lies: lies, damned lies and statistics"). The second and third refer briefly to the standards of measure used in the book. The fourth note, which is the first citation in chapter 1, is to the Worldwatch Institute's annual publication, *The State of the World*, which points out that Lester Brown "was president for Worldwatch Institute till 2000." The fifth note reports that there are better publications than *The State of the World* "from an academic point of view," including publications from the United Nations, "as well as all the fundamental research, much of which is used in this book and can be found in the bibliography."

As Lomborg begins to describe how the "Litany" of bad environmental news is communicated to the public, he cites *Time* magazine (note 6), a children's book (note 7), *Time* magazine again (note 8), and an article from *New Scientist* (note 9). *Time* had reported, as recounted by Lomborg, that "everyone knows the planet is in bad shape," and "for more than 40 years, earth has been sending out distress signals." A children's book had also reported: "The balance of nature is delicate but essential for life. Humans have upset that balance, stripping the land of its green cover, choking the air, and poisoning the seas." And *New Scientist* "talks about the impending 'catastrophe' and how we risk consigning 'humanity to the dustbin of evolutionary history.' "

In note 10 Lomborg relates that his use of the word "Litany" (that is, the bad news about the environment reported by the environmental organizations) comes from a 1997 article in *Wired* magazine. Note 11 catches *Time* magazine again imparting the Litany. Note 12 cites Isaac Asimov and Frederik Pohl also preaching the Litany. After identifying the environmental "fundamentals" that Lomborg is about to review ("the number of people on earth," "air pollution," "global warming," "chemical fears," and "pesticides"), note 13 comments: "It is impossible to cover *all* important areas, but I believe that this book covers most of them. . . . New suggestions, of course, are always welcome."[4]

Up to this point (the first two pages of chapter 1) Lomborg has logged thirteen endnotes, the most substantive of which document what he views as rhetorical excesses by journalists and environmentalists; none yet to document his thesis that such excesses reflect a skewed perspective of the *real* global environment. He seems to begin to address this with his fourteenth endnote, which follows his statement, "We are not running out of energy or natural resources," and which reads (note 14): "This and the following claims are documented in the individual chapters below."[5] Lomborg's opportunity to support his thesis occurs immediately after note 14, when he issues a series of highly substantive claims about the *real* state of the world's environment, yet he fails to document a single claim:

> There will be more and more food per head of the world's population. Fewer and fewer people are starving. In 1900 we lived for an average of 30 years; today we live for 67. According to the UN we have reduced poverty more in the last 50 years than we did in the preceding 500, and it has been reduced in practically every country.
>
> Global warming, though its size and future projections are rather unrealistically pessimistic, is almost certainly taking place, but the typical cure of early and radical fossil fuel cutbacks is way worse than the original affliction, and moreover its total impact will not pose a devastating problem for our future. Nor will we lose 25–50 percent of all species in our lifetime—in fact we are losing probably 0.7 percent. Acid rain does not kill the forests, and the air and water around us are becoming less and less polluted.
>
> Mankind's lot has actually improved in terms of practically every measurable indicator.[6]

While life expectancy has increased, and the condition of major segments of humankind has improved, environmentalists point to unaccounted-for environmental costs and impacts.[7] Furthermore, when Lomborg argues that the air and water are less polluted, he refers primarily to developed Western states, and neglects to mention that cleaner air and water are due to the kind of government regulations that he now opposes with respect to global warming and greenhouse emissions. As for the other claims—that fewer people are starving, there will be more food for people in the future, the threat of global warming is exaggerated, cutting fossil fuel consumption would be worse than the effects of global warming, the rate of species extinction is very small, and acid rain is not harmful—Lomborg presents these assertions without endnotes or otherwise citing any data, nor does he provide convincing data later in the book as he promised in note 14.

For example, Lomborg never provides published peer-reviewed data to support his claim that we will lose only 0.7 percent of all species throughout our lifetime. For one thing, "throughout our lifetime" is an unspecified period of time, and, for another, a 0.7 percent loss of species is a highly precise projection. It is unlikely that any studies exist that reconcile these seemingly incompatible variables.

Though Lomborg did not document these factual assertions, which are central to the thesis of his book, the next citation (note 15) references the distinction between *is* and *ought* from David Hume's *A Treatise of Human Nature*, published in 1740.[8]

Furthermore, the historical trends that Lomborg uses, such as those cited above concerning improvements in life expectancy, are not, by themselves, sufficient to evaluate what may happen in the future. Yet much of Lomborg's analysis, including his environmental "fundamentals," is backward-looking, as he notes: "If we are to understand the real state of the world, we need to focus on the *fundamentals* and we need to look at *realities*, not myths. Let us take a look at both of these requirements, starting with the fundamentals. When we are to assess the state of the world, we need to do so through a comparison. Legend has it that when someone remarked to Voltaire, 'life is hard,' he retorted, 'compared to what?' Basically, the choice of comparison is crucial. It is my argument that the comparison should be with *how it was before*. Such comparison shows us the extent of our progress—are we better or worse off now than previously? This means that we should focus on *trends*"[9] (emphasis in original).

Not only is this level of analysis, while helpful in many contexts, not very useful with respect to assessing the future impact of the relatively recent phenomenon of global warming, but Lomborg fails to avoid some pitfalls, such as neglecting the consideration of latent but identifiable underlying costs and conditions. For example, if homeownership is higher in 2005 than it was in 1990, Lomborg might argue that the *trend* in homeownership is positive; but if he ignores conditions that portend a mortgage-foreclosure calamity, and an ensuing financial and economic crisis, then the cited trend can be misleading as an indicator of future well-being.

This is what Lomborg largely does in the context of the world's environment and global warming. A case in point is the following statement by Lomborg in *The Skeptical Environmentalist:* "The point is that ever fewer people in the world are starving. In 1970, 35 percent of all people in developing countries were starving. In 1996 the figure was 18 percent and the UN expects that the figure will have fallen to

12 percent by 2010. This is remarkable progress: 237 million fewer people starving. Till today, more than 2 billion more people are getting enough to eat. The food situation has vastly improved, but in 2010 there will still be 680 million people starving, which is obviously not *good enough*. . . . But when things are improving we know we are on the right track."[10]

Lomborg sourced the statistical percentages of this statement (endnote 16) up to the sentence ending, "fallen to 12 percent by 2010," but the subsequent assertions are not supported with documentation. Endnote 16 cites two tables: one from a 1996 World Food Summit report issued by the UN's Food and Agriculture Organization (FAO),[11] and another from a 1999 FAO report.[12] While Lomborg wrote that these reports demonstrate "remarkable progress," "that we are on the right track," but that results are "obviously not good enough," the 1999 FAO report observes that these numbers "fall squarely within the trajectory of 'business as usual,'" and that this "state of affairs is unacceptable."[13] Nor did Lomborg mention another observation from the 1999 FAO report: "A closer look at the data reveals that in the first half of this decade a group of only 37 countries achieved reductions [in malnutrition] totaling 100 million [people]. Across the rest of the developing world, the number of hungry people actually increased by almost 60 million."[14]

Nor did Lomborg mention that the 1999 report stated that "floods, drought, war and financial collapse threaten progress."[15] Nor did the 1999 report consider or mention "global warming" or "climate change," which obviously reduces its value as a prognostication tool for future worldwide malnutrition in the context of global warming. And though Lomborg noted that the reported progress in reducing malnutrition "is obviously not good enough," there were few indications in the FAO report that "we are on the right track," as Lomborg commented.

Indeed, in the 2004 "The State of Food Insecurity," the FAO reported: "FAO estimates that 852 million people worldwide were undernourished in 2000–2002. This figure includes 815 million in developing countries, 28 million in the countries in transition and

9 million in the industrialized countries. The number of undernourished people in developing countries decreased by only 9 million during the decade following the World Food Summit baseline period of 1990–92. During the second half of the decade, the number of chronically hungry in developing countries increased at a rate of almost 4 million per year, wiping out two thirds of the reduction of 27 million achieved during the previous five years."[16]

Though Lomborg cannot be held responsible for failing to report this increase in world malnutrition (it was reported after *The Skeptical Environmentalist* was published), it is apparent that his assessment of "remarkable progress" in reducing malnutrition was at best premature. And though Lomborg overstated progress in reducing malnutrition—he also wrote that we "can look forward to fewer people starving in future,"[17] a projection not supported by the 2004 FAO report—he drew an important lesson upon claiming such progress, which exemplifies Lomborg's approach: "The constant repetition of the Litany [in this case, that too many people are malnourished] and the often heard environmental exaggerations has serious consequences. It makes us scared and it makes us more likely to spend our resources and attention solving phantom problems while ignoring real and pressing (possibly non-environmental) issues. This is why it is important to know the real state of the world. We need to get the facts and the best possible information to make the best possible decisions."[18] But in this case and others, the Lomborg-asserted "best possible information" was not supported by his endnotes.

Continuing to build one unsubstantiated assertion upon another, Lomborg sought to affiliate his efforts with Gro Harlem Brundtland—a former prime minister of Norway and distinguished physician, public health expert, and environmentalist: "As the lead author of the environmental report *Our Common Future*, Gro Harlem Brundtland, put it in the top scientific magazine *Science:* 'Politics that disregard science and knowledge will not stand the test of time. Indeed, there is no other basis for sound political decisions than the best available scientific evidence. This is especially true in the fields of resource management and environmental protection.' "[19] These comments, however, presented in this context, are meaningless, since Lomborg's book doesn't conform to the principles espoused; however, they are potentially effective as a literary device by giving an impression that Lomborg's "things are getting better" argument is supported by a principled advocate of science, that is, himself. And with respect to the endnote count, Lomborg's citation for Gro Harlem Brundtland was note 17. This means that up to this point Lomborg provided little evidence of his thesis that the major environmental organizations peddle an unsupported Litany of environmental damage and that the global environment is well off.

Endnote 18 followed the statement "When we are to assess the state of the world, we need to do so through a comparison." This statement reinforced Lomborg's notion about how we should forecast our environmental future by comparing things today with how they were before.

Note 19 supported the following sentence: "Legend has it that when someone remarked to Voltaire, 'life is hard,' he retorted, 'compared to what?' "

Note 20 supported the uncontroversial statement: "When the water supply and sanitation services were improved in cities throughout the developed world in the nineteenth century, health and life expectancy improved dramatically."

Note 21 supported the statement that "universal school enrollment has brought literacy and democratic competence to the developed world." These are among the historical trends that Lomborg invoked in this section.[20]

Referring to increased access to drinking water and sanitation (sewage systems) in the developing world, and to improvements in literacy, Lomborg wrote (the citations in parentheses are in the original): "These trends have been replicated in the developing world in the twentieth century. Whereas 75 percent of the young people in the developing world born around 1915 were illiterate, this is true for only 16 percent of today's youth (see Figure 41, p. 81). And while only 30 percent of the people in the developing world had access to clean drinking water in 1970, today about 80 percent have (see Figure 5,

p. 22)."[21] Lomborg attributes Figure 41 to page 8 in a 1990 UNESCO document, but the figure cannot be located in the document.[22] And Lomborg himself drew Figure 5, which he described as "a reasonable attempt to map out the best guess of development among very different definitions" of access to drinking water and sanitation.[23]

Lomborg also wrote that "women still do not have the same access to education, and this is also reflected in the higher illiteracy rate, which at 21 percent is almost double that of men at 12 percent."[24] Lomborg supports this assertion by referencing it to a 1998 UNESCO document—"Gender-Sensitive Education Statistics and Indicators"—that cannot be found using the Lomborg-provided URL or document title.[25] However, a 1997 UNESCO document with a nearly identical title—"Gender Sensitive Education Statistics and Indicators: A Practical Guide"—reports that the illiteracy rate in the developing world in 1995 was 38 percent among women and 21 percent among men,[26] not the 21 percent among women and 12 percent among men that Lomborg reported. In addition, the 1997 UNESCO document reported a 53 percent illiteracy rate among women in Sub-Saharan Africa (33 percent for men), 56 percent among women in Arab states (32 percent for men), and 63 percent for women in southern Asia (37 percent for men).[27]

Following his comments about malnutrition, literacy, and access to drinking water and sanitation, Lomborg began a section titled, "Fundamentals: Global Trends." The first sentence reads: "The *Global Environment Outlook Report 2000* tells us much about the plight of Africa." Endnote 22 (and its corresponding reference in the bibliography) gives the full title and URL of this UN-based report. Lomborg then wrote: "Sub-Saharan Africa has by far the greatest numbers of starving people—almost 33 percent were starving in 1996, although this was down from 38 percent in 1970 and is expected to fall even further to 30 percent in 2010."[28] Endnote 23 provides two sources to document Lomborg's claim that Sub-Saharan Africa "has by far the greatest numbers of starving people," and that this total is "down" and "is expected to fall further." The first source is the FAO's 1996 "World

Food Summit" report,[29] and the second source is the FAO's 1999 publication *The State of Food Insecurity in the World.*[30] However, neither of these reports uses the word "starving" in its text or charts; they use "under-nutrition" and "undernourished" for the "starving" category that Lomborg cites — the word "starving" does not appear in either of the reports. Also, both of these FAO reports indicate that South Asia (consisting of India, Bangladesh, Nepal, Pakistan, and Sri Lanka) and not Sub-Saharan Africa has "the greatest numbers" of undernourished persons, though the percentage of undernourished persons in South Asia is smaller than the percentage in Sub-Saharan Africa.[31] The 1999 FAO report, which gives more recent statistics, stipulates that the number of undernourished persons in South Asia from 1995 to 1997 was 284 million, and that the number for the same period in Sub-Saharan Africa was 180 million.[32] The 1999 FAO report, in fact, plainly states: "India alone has more undernourished people (204 million) than all of Sub-Saharan Africa combined."[33]

Though Lomborg claimed that the *percentage* of "starving people" in Sub-Saharan Africa is "down" and "is expected to fall even further," both the 1996 and 1999 FAO reports indicate that the *numbers* of undernourished people in Sub-Saharan Africa have generally increased. For example, Table 3 of the 1996 FAO report indicates that the number of undernourished persons in Sub-Saharan Africa in 1969–71 was 103 million, the number in 1979–81 was 148 million, and the number in 1990–92 was 215 million.[34] The 1999 FAO report — *The State of Food Insecurity in the World*, the first of several annual reports by the same title — indicates that there were 180 million undernourished persons in Sub-Saharan Africa in 1995–97, a decrease from the previous five-year period, but not within the thirty-year period reported.[35]

Subsequent FAO reports issued after *The Skeptical Environmentalist* was published in 2001 show further increases in the number of undernourished persons in Sub-Saharan Africa, with no appreciable reduction in the percentage of undernourished persons: 180 million undernourished in Sub-Saharan Africa in 1996 (as reported in the 1999 FAO report) and 206 million undernourished in 2002 (as reported in the 2006 FAO report); and 33 percent undernourished in 1996 (as reported in the 1999 report) with 32 percent undernourished in 2002 (as reported in the 2006 report). In fact, the 2006 FAO report stated: "Hunger in sub-Saharan Africa is as persistent as it is widespread. Between 1990–92 and 2001–03, the number of undernourished people increased from 169 million to 206 million, and only 15 of the 39 countries for which data are reported reduced the number of undernourished."[36]

Even though Lomborg argued, while citing the 1996 World Food Summit report, that the percentage of "starving people" in Sub-Saharan Africa "is expected to fall even further to 30 percent in 2010," the same line of the same table (Table 3) in the same report from which Lomborg derived that percentage also projected that the number of undernourished people in Sub-Saharan Africa by 2010 would increase to 264 million (a figure that Lomborg ignored) — which represents an increase of forty-nine million undernourished persons from 1990 to 1992 (per Table 3), and an increase of 161 million undernourished persons from 1969 to 1971 (per Table 3).[37]

Lomborg not only contradicts his own documented sources when he claims that malnutrition is down significantly in Sub-Saharan Africa and is expected to drop further, he sanitizes the human impact of the state of malnutrition by ignoring other key statistics. For example, the 1999 FAO report noted that "two out of five children in the developing world are stunted, one in three is underweight and one in ten is wasted."[38] Furthermore, a country-by-country analysis, in addition to the broader regional analysis used by Lomborg, shows that the percentage of undernourished persons is above 33 percent of the total population in nearly half the countries of the Sub-Saharan region, including: 26 million undernourished persons in a total population of 47 million (55 percent) in the Democratic Republic of Congo; 4 million undernourished in a population of 6.3 million (63 percent) in Burundi; 2.2 million undernourished in a population of 3.3 million (67 percent) in Eritrea; 29 million undernourished in a population of 57 million (51 percent) in Ethiopia; 6.2 million undernourished in a population of 8.5 million (73 percent) in Somalia; and 11.3 million undernourished in a population of 17.9 million (63 percent) in Mozambique.[39] Overall, Lomborg's claim that malnutrition has improved in Sub-Saharan Africa is not supported by the sources that he cited.

Lomborg proceeded to argue in a comparably flawed manner that the degree of soil degradation in Africa—a key factor underlying the problem of undernourished African populations—is also exaggerated. He wrote: "In the most staggering prediction of problems ahead, *Global Environmental Outlook Report 2000* tells us that soil erosion is a pervasive problem, especially in Africa. Indeed, 'in a continent where too many people are already malnourished, crop yields could be cut in half within 40 years if the degradation of cultivated lands were to continue at present rates.' This, of course, would represent a tragedy of enormous proportions, causing massive starvation on the African continent. However, the background for this stunning prediction stems from a single, unpublished study from 1989, based on agricultural plot studies only in South Africa."[40] There is little evidence to support Lomborg's general claim that the issue of soil degradation in Africa is overblown. For example, in endnote 24 Lomborg cited the *GEO-2000* report itself to support his claim that *GEO-2000* used only one unpublished source while arguing that soil degradation in Africa is such a serious environmental issue that it could lead to a 50 percent reduction in crop yields in Africa within forty years. However, the relevant page of *GEO-2000* cited several authoritative sources while describing serious land degradation in Africa:

> Land degradation is a serious problem throughout Africa, threatening economic and physical survival. Key issues include escalating soil erosion, declining fertility, salinization, soil compaction, agrochemical pollution and desertification. An estimated 500 million hectares of land have been affected by soil degradation since about 1950 (UNEP/ISRIC 1991), including as much as 65 per cent of agricultural land (Oldeman 1994). Soil losses in South Africa alone are estimated to be as high as 400 million tonnes annually (SARDC, IUCN and SADC 1994). Soil erosion affects other economic sectors such as energy and water supply. *In a continent where too many people are already malnourished, crop yields could be*

> *cut by half within 40 years if the degradation of cultivated lands were to continue at present rates* (Scotney and Dijkhuis 1989).[41] [Emphasis added]

Although Lomborg correctly asserted that the one sentence (italicized above) he extracted from this paragraph was supported by one source, he neglected to mention that the overall paragraph in which the sentence appeared included several statements supported by several sources describing serious land degradation in Africa. Furthermore, the one source to which Lomborg referred (Scotney and Dijkhuis 1989), which he described as "an unpublished study from 1989," was in fact published in the *South African Journal of Science* ("Changes in the Fertility Status of South African Soils," 1990: *S. Afr. J. Sci.* 86:395–402), which is issued by the Academy of Science of South Africa. In endnote 25 Lomborg wrote: "Despite several attempts, I was unable to get hold of this publication." I telephoned the reference library at the University of Massachusetts at Amherst, where a librarian located the study in only a few minutes.

Referring to the warning in *GEO-2000* that crop yields in Africa could be cut by half in forty years if the problem of soil degradation is not addressed, Lomborg wrote that this assessment "is in stark opposition to the estimates of the major food production models from the UN (FAO) and IFPRI [International Food Policy Research Institute], expecting an annual 1.7 percent yield increase over the next 20–25 years."[42] Endnote 26 identified the FAO and IFPRI reports. But there is no "stark opposition" in the FAO and IFPRI reports to the warning in *GEO-2000*.[43] For example, the more recently issued IFPRI report (1999), like *GEO 2000*, described serious problems with soil fertility in Africa: "Low and declining soil fertility is a serious problem in many low-income countries, including most of Africa."[44] And with respect to the future of undernourished persons in the developing world, including those in Sub-Saharan Africa, the 1999 IFPRI report also stated: "In the scenario described here, food insecurity and malnutrition will persist in 2020 and beyond. We project that 135 million children under five years of age will be malnourished in 2020, a decline of only 15 percent from 160 million in 1995. Child malnutrition is expected to decline in all major developing regions except Sub-Saharan Africa, where the number of malnourished children is forecast to increase by about 30 percent to reach 40 million by 2020. With more than 77 percent of the developing world's malnourished children in 2020, Sub-Saharan Africa and South Asia will remain 'hot spots' of child malnutrition and food insecurity."[45]

It is difficult to find here any good news about the future of undernourished persons, especially in Sub-Saharan Africa. Furthermore, while this detailed report projects broad incremental increases in income levels in the developing world from 1995 to 2020, it would be misleading to characterize these statistics in many cases as significant progress, and thus projecting any serious steps forward in reducing malnutrition. For example, the 1999 IFPRI report states that income in Sub-Saharan Africa will increase at a 3.4 percent annual rate from 1995 to 2020; however, this will increase per capita annual income from US$280 per person in 1995 in Sub-Saharan Africa to only US$359 per person in 2020. Likewise, IFPRI reports that income will rise in South Asia at an annual rate of 5 percent from 1995 to 2020; however, this will increase per capita annual income from US$350 per person in 1995 in South Asia to US$830 per person in 2020. Overall, income is expected to increase throughout the developing world at a rate of 2.2 percent from 1995 to 2020, increasing average per capital annual income from US$1,080 in 1995 to US$2,217 in 2020. While this would double incomes from 1995 to 2020, by 2020 average income in the poorer developing world would still be fourteen times lower than in the richer developed world.[46]

Lomborg argues throughout that such trends, expressed as broad averages and percentages, reflect progress when he writes that "as we become richer" this century—including "developing countries as they, too, get ever richer"—our standards of living will improve.[47] However, such trends often reflect at best "the Bill Gates effect," as described by *New York Times* columnist Paul Krugman: "Averages can be deeply misleading. When Bill Gates enters a bar, the average net worth of the patrons soars, but that doesn't make everyone in the bar a billionaire."[48] Lomborg's frequent use of broad statistical averages to claim progress in a given area of development likewise often makes "billionaires" out of millions of the world's poorest people, who lack access to food, clean water, and sanitation. Lomborg would have done better, with respect to malnutrition and throughout, to report a fuller range of indicators beyond broad percentages when citing alleged progress so as to permit readers to better assess the extent (or lack) of progress.

Finally, after Lomborg implied that the *GEO-2000* assessment of soil degradation in Sub-Saharan Africa was grounded in one unpublished study, when in fact the study in question was published and *GEO-2000* used several sources to describe soil degradation in Africa, and after he argued that *GEO-2000* is in "stark opposition" to a 1995 FAO report and a 1999 IFPRI report, when in fact no significant tensions exist between these two reports and *GEO-2000*, he wrote in that same section and paragraph: "Although the growth in [agricultural] yield in the 1990s was small but positive, the absolute grain production increased more than 20 percent."[49] However, there is no indication in this sentence, or in the endnote to this sentence (note 27), whether "the growth in yield in the 1990s" and the increase in "absolute grain production" refers to Sub-Saharan Africa, Africa in general, the developing world, the developed world, the *real* word, or the *entire* world. And Lomborg's endnote to this assertion refers to a document for which Lomborg gives no title, and which cannot be located on the Internet using the URL that Lomborg provided.[50] Thus, it is Lomborg (on at least a few counts) — and not *GEO 2000* — who made unsubstantiated assertions.

Though Lomborg provided no actual evidence to support his charge that the *GEO-2000* report exaggerated the issue of soil degradation and malnutrition in Africa, he proceeded as if his criticism of *GEO-2000* laid the foundation for more such charges of environmentalists' exaggerations. Thus, immediately following his comments on *GEO-2000*, Lomborg wrote: "In many ways this [the alleged *GEO-2000* exaggeration] is reminiscent of one of the most cited European soil erosion estimates of 17 tons per hectare."[51] The word "this" — a word that Lomborg uses often and whose antecedents are often unclear—refers to Lomborg's apparent belief that he had successfully impeached the analysis in *GEO-2000* pertaining to soil degradation in Africa, and that the next alleged exaggeration (pertaining to soil erosion in Europe) can be stacked solidly upon *GEO-2000's* exaggerations about soil degradation in Africa.

In this instance, Lomborg wrote: "This estimate [that 17 tons of soil per hectare are eroding] turned out—through a string of articles, each slightly inaccurately referring to its predecessor—to stem from a single study of a 0.11 hectare sloping plot of Belgian farmland, from which the author himself warns against generalization."[52] This is the second time within five sentences in *The Skeptical Environmentalist* where Lomborg accused environmental researchers of using a lone, dubious source as evidence to support a description of serious environmental damage.

Though Lomborg doesn't mention these details in the text of his book, the target of his accusation in this instance, as detailed in his notes and bibliography sections, is a study published in *Science* by David Pimentel, a prominent scientist at Cornell's College of Agriculture and Life Sciences and an expert on environmental science, biodiversity, and soil conservation.[53] Putting aside what Lomborg's allegations might imply about his views of the publication standards at *Science* and of a top scientist at arguably the best school of agriculture in the United States, Lomborg wrote nothing more about Pimentel or his paper beyond the two sentences quoted above. And since Lomborg provided no time framework for the loss of 17 tons per hectare, as he presented it, it is not immediately clear whether this loss of soil is per year, per decade, or per century.[54] Nevertheless, in the next few sentences, Lomborg explicitly tied the alleged *GEO-2000* exaggeration about soil erosion in Africa to the alleged Pimentel/*Science* exaggeration about soil erosion in Europe, and commented: "In both examples, sweeping statements are made with just a single example. Unfortunately, such problematic argumentation is pervasive, and we will see more examples below. The problem arises because in today's global environment, with massive amounts of information at our fingertips, an infinite number of stories can be told, good ones and bad."[55]

Like Lomborg's inaccurate claim that *GEO-2000* had relied upon a single unpublished source while exaggerating the extent of soil degradation in Africa, the allegations here against Pimentel's paper in *Science* are also misleading. Here is the context, including the comparative context, in which Pimentel presented the 17 tons of soil erosion estimate: "Soil erosion rates are highest in Asia, Africa, and South America, averaging 30 to 40 tons [per hectare per] year, and lowest in the United States and Europe, averaging about 17 tons [per hectare per] year."[56] Without mentioning that Pimentel and his colleagues had reported that the United States and Europe had the lowest rates of soil erosion in the world, Lomborg implied that Pimentel had exaggerated estimates of soil erosion in Europe. Even so, and more importantly, in the next sentence of the study in *Science*, Pimentel wrote: "The relatively low rates of soil erosion in the United States and Europe, however, greatly exceed the average rate of soil formation of about 1 ton [per hectare per year]."[57] This means that even if Pimentel had exaggerated soil erosion rates in Europe, say, by doubling the actual erosion rate, this would still leave an undesirable eight-to-one ratio of soil erosion to soil formation. Thus, one might ask why Lomborg would leave these key portions of Pimentel's paper out of *The Skeptical Environmentalist*, even as he accused Pimentel of exaggerating soil erosion rates in Europe.

Furthermore, Lomborg's allegations against Pimentel and colleagues, and by extension against one of the world's most prestigious science journals, divert the reader from the overall point and larger body of evidence in Pimentel's paper, which undermine Lomborg's contention that "things are getting better" with respect to the ability of the Earth to feed a rapidly growing human population. The first two paragraphs of Pimentel's study describe the fundamental problem, which Lomborg ignored (I substituted "hectare" below for its abbreviation "ha" in the original):

> Soil erosion is a major environmental and agricultural problem worldwide. Although erosion has occurred throughout the history of agriculture, it has intensified in recent years. Each year 75 billion metric tons of soil are removed from the land by wind and water erosion, with most

> coming from agricultural land. The loss of soil degrades arable land and eventually renders it unproductive. Worldwide, about 12×10^6 hectares of arable land are destroyed and abandoned annually because of nonsustainable farming practices, and only about 1.5×10^9 hectares of land are being cultivated. Per capita shortages of arable land exist in Africa, Asia, and Europe because of lost eroded land and the expansion of the world population to nearly 6 billion.
>
> To adequately feed people a diverse diet, about 0.5 hectares of arable land per capita is needed, yet only 0.27 hectares per capita is available. In 40 years, only 0.14 hectares per capita will be available both because of loss of land and rapid population growth. In many regions, limited land is a major cause of food shortages and undernutrition. Over 1 billion humans (about 20% of the population) now are malnourished because of food shortages and inadequate distribution. With the world population increasing at a quarter of a million per day and continued land degradation by erosion, food shortages and malnutrition have the potential to intensify.[58]

For the record, Pimentel's two-paragraph summary (above) was supported with 11 documented sources. Yet this crucial information pertaining to shrinking agricultural land worldwide juxtaposed with a growing human population with growing food needs, authored by leading experts and published in a leading science journal, never found its way into Lomborg's *The Skeptical Environmentalist.*

In this chapter I have used about 9,000 words to review the first twenty-nine endnotes in Lomborg's *The Skeptical Environmentalist* (including Lomborg's references to Figures 5 and 41 on pages 22 and 82). To similarly review all of the 2,930 endnotes in Lomborg's book would require a 900,000-word book comprising more than 100 chapters the size of this one. Therefore, if one were to assess the supporting evidence and accuracy of "Lomborg's Theorem" (that global warming is "no catastrophe") and "Lomborg's Corollary" (that we therefore should not prioritize the reduction of greenhouse emissions), reviewing the entire text of *The Skeptical Environmentalist* would not be the most practical approach to achieving that goal. That book, published in 2001, is also focused on other environmental issues in addition to global warming. In contrast, Lomborg's 2007 book, *Cool It*, is smaller, is focused exclusively on global warming, was published more recently, and updates Lomborg's Theorem and Lomborg's Corollary as he first presented them in *The Skeptical Environmentalist.* Thus, reviewing *Cool It* — as a means of reviewing Lomborg's work about global warming — is both more relevant and practical.

Up to this point I have sought to give the reader insight into Lomborg's literary methods from, first, the perspective of several prominent scientists and environmentalists as it pertained to their readings of *The Skeptical Environmentalist;* second, the perspective of my own analysis of Lomborg's first chapter in *Cool It* on polar bears as a case study on how environmentalists allegedly exaggerate the larger issue of global warming; and third, a review of how Lomborg utilized the first 1 percent of his endnotes in *The Skeptical Environmentalist.* It is my hope that these chapters will provide the context for a more focused analysis of Lomborg's Theorem as he has more recently presented it in *Cool It,* and as I review the theorem in the following chapters with a close reading of that book.

PART 2

Lomborg's Theorem

four
GLOBAL WARMING IS "NO CATASTROPHE"

The cumulative impact of the first three chapters introduces the possibility that Lomborg's methodology is chronically flawed, and that Lomborg's Theorem (that global warming is "no catastrophe") and Lomborg's Corollary (that we can thus reject substantial reductions in greenhouse emissions) are grounded, to say the least, in bad data. In the next several chapters I review Lomborg's Theorem as presented in his 2007 book *Cool It*, so as to determine the fuller extent of Lomborg's data problems as applied to his updated and focused analysis of the threat of global warming. Because I've already reviewed *Cool It*'s first chapter, "On Polar Bears," I begin with chapter 2, "It's Getting Hotter: The Short Story."

Lomborg argues in *Cool It* that, on balance, an increase in global temperatures this century due to greenhouse emissions would threaten "no catastrophe" for the Earth and its species of plants and animals, including humans.[1] In making this argument, with a few minor exceptions, Lomborg avoided citing the scientific benchmark on this question—the global-warming assessment reports of the UN's Intergovernmental Panel on Climate Change (IPCC). In its twenty-year tenure the IPCC has issued four such reports to date—in 1990, 1995, 2001, and 2007. Hereinafter I will refer to a group of two or more of these reports as the "IPCC assessment reports," or the "IPCC assessments," or to a specific report by year ("1990 IPCC assessment report," "2001 IPCC assessment," etc.).

These assessment reports, especially the 2001 and 2007 IPCC assessments, should have been the baseline sources for Lomborg's depictions of climate change impacts, especially since he argued that there would be few (if any) harmful impacts, while the IPCC had described and projected many harmful impacts. For example, the IPCC's 2007 assessment report, which Lomborg cited infrequently in

Cool It, projected severe regional impacts of global warming, including the ones summarized below:

Africa

- By 2020, between 75 and 250 million people are projected to be exposed to increased water stress due to climate change.
- Agricultural production, including access to food, in many African countries is projected to be severely compromised by climate variability and change.
- Local food supplies are projected to be negatively affected by decreasing fisheries resources in large lakes due to rising water temperatures.

Asia

- Freshwater availability in Central, South, East and South-East Asia, particularly in large river basins, is projected to decrease due to climate change which, along with population growth and increasing demand arising from higher standards of living, could adversely affect more than a billion people by the 2050s.
- Climate change is projected to impinge on the sustainable development of most developing countries of Asia as it compounds the pressures on natural resources and the environment associated with rapid urbanization, industrialization and economic development.
- Endemic morbidity and mortality due to diarrhea-related disease primarily associated with floods and droughts are expected to rise in East, South and South-East Asia due to projected changes in the hydrological cycle associated with global warming. Increases in coastal water temperatures would exacerbate the abundance and/or toxicity of cholera in South Asia.

Australia and New Zealand

- As a result of reduced precipitation and increased evaporation, water security problems are projected to intensify by 2030 in southern and eastern Australia and, in New Zealand, in Northland and some eastern regions.
- Significant loss of biodiversity is projected to occur by 2020 in some ecologically rich areas, including the Great Barrier Reef and Queensland Wet Tropics.
- Production from agriculture and forestry by 2030 is projected to decline over much of southern and eastern Australia, and over

parts of eastern New Zealand, due to increased drought and fire. However, initially in New Zealand, agricultural benefits are projected in western and southern areas.

Europe

- For the first time, wide-ranging impacts of changes in current climate have been documented: retreating glaciers, longer growing seasons, shift of species ranges, and health impacts due to a heat wave in 2003 of unprecedented magnitude. The observed changes described above are consistent with those projected for future climate change.
- Nearly all European regions are anticipated to be negatively affected by some future impacts of climate change, and these will pose challenges to many economic sectors. Negative impacts will include increased risk of inland flash floods, and more frequent coastal flooding and increased erosion (due to storminess and sea-level rise). The great majority of organisms and ecosystems will have difficulty adapting to climate change. Mountainous areas will face glacier retreat, reduced snow cover and winter tourism, and extensive species losses (in some areas up to 60% under high emission scenarios by 2080).

Latin America

- By mid-century, increases in temperatures and associated decreases in soil water are projected to lead to gradual replacement of tropical forest by savanna in eastern Amazonia. Semi-arid vegetation will tend to be replaced by arid-land vegetation. There is a risk of significant biodiversity loss through species extinction in many areas of tropical Latin America.
- In drier areas, climate change is expected to lead to salinization and desertification of agricultural land. Productivity of some important crops is projected to decrease and livestock productivity to decline, with adverse consequences for food security. However, in temperate zones soybean yields are projected to increase.
- Changes in precipitation patterns and the disappearance of glaciers are projected to significantly affect water availability for human consumption, agriculture and energy generation.

North America

- Warming in western mountains is projected to cause decreased snowpack, more winter flooding, and reduced summer flows, exacerbating competition for over-allocated water resources.

- Disturbances from pests, disease and fire are projected to have increasing impacts on forests, with an extended period of high fire risk and large increase in area burned.
- Moderate climate change in the early decades of the century is projected to increase aggregate yields of rain-fed agriculture by 5–20 percent, but with variability among regions. Major challenges are projected for crops that are near the warm end of their suitable range or which depend on highly utilized water resources.
- Cities that currently experience heat waves are expected to be further challenged by an increased number, intensity and duration of heat waves during the course of the century, with potential for adverse health impacts. Elderly populations are most at risk.
- Coastal communities and habitats will be increasingly stressed by climate change impacts interacting with development and pollution. Population growth and the rising value of infrastructure in coastal areas increase vulnerability to climate variability and future climate change, with losses projected to increase if the intensity of tropical storms increases.

Polar Regions

- The main projected biophysical effects are reductions in thickness and extent of glaciers and ice sheets, and changes in natural ecosystems with detrimental effects on many organisms including migratory birds, mammals and higher predators. In the Arctic, additional impacts include reductions in the extent of sea ice and permafrost, increased coastal erosion, and an increase in the depth of permafrost seasonal thawing.
- For human communities in the Arctic, impacts, particularly those resulting from changing snow and ice conditions, are projected to be mixed. Detrimental impacts would include those on infrastructure and traditional indigenous ways of life. Beneficial impacts would include reduced heating costs and more navigable northern sea routes.
- In both polar regions, specific ecosystems and habitats are projected to be vulnerable, as climatic barriers to species invasions are lowered.

Small Islands

- Deterioration in coastal conditions, for example through erosion of beaches and coral bleaching, is expected to affect local resources, e.g., fisheries, and reduce the value of these destinations for tourism.

- Sea level rise is expected to exacerbate inundation, storm surge, erosion and other coastal hazards, thus threatening vital infrastructure, settlements and facilities that support the livelihood of island communities. Climate change is projected by mid-century to reduce water resources in many small islands, e.g., in the Caribbean and Pacific, to the point where they become insufficient to meet demand during low-rainfall periods.[2]

It is telling that Lomborg would ignore these matter-of-fact projections of major environmental damage and threats while filling *Cool It* with dubious accusations that environmentalists exaggerate the harmful effects of global warming.

Rather than engage these findings, Lomborg instead inserted eleven endnotes in chapter 2 on the IPCC's work,[3] mostly to support preliminary or mundane observations, and without addressing the IPCC's projected impacts. For example, Lomborg briefly noted that the so-called greenhouse effect—independent of human-induced influences—is beneficial and necessary for life on Earth as we know it: "Several types of gases can reflect or trap heat, most importantly water vapor and carbon dioxide (CO_2). These greenhouse gases trap some of the heat emitted by the Earth, rather like a blanket wrapped around the globe. The basic greenhouse effect is good: if the atmosphere did not contain greenhouse gases, the average temperature on the Earth would be approximately 59°F colder, and it is unlikely that life as we know it would be able to exist. The problem is that people have substantially increased the quantity of CO_2 in the atmosphere, mainly from burning fossil fuels, such as coal, oil, and gas. As natural processes only slowly remove CO_2 from the atmosphere, our annual emissions have increased the total atmospheric CO_2 content—the CO_2 concentration—such that today it is 36 percent higher than in preindustrial times."[4]

Lomborg then wrote: "Absent a major policy change, we will continue to burn more fossil fuels over the coming century. This is especially true for the rapidly industrializing developing world, such as China and India. Whereas the developing world now is responsible for about 40 percent of the annual global carbon emissions, by the end of the century that figure will more likely be 75 percent. More CO_2 will hold on to more heat and raise temperatures. This is the man-made greenhouse effect."[5] To my knowledge, this is an acceptable introductory summary of the greenhouse effect and human-induced global warming, which Lomborg supported by citing four of the eleven IPCC sources that he used in the chapter.[6]

Lomborg continued: "Let's look at what will happen when we turn up the heat," acknowledging that "our best information" on the projected impact of human-induced climate warming "comes from the United Nations' Intergovernmental Panel on Climate Change, or IPCC." Lomborg noted that "every six years or so, [the IPCC] gathers the best information we have on climate models and climate effects."[7] Thus, up to this point, Lomborg reasonably summarized the nature of human-induced global warming, and acknowledged that overall the IPCC provides the benchmark science on climate change and its projected impacts. Things fall apart, however, when Lomborg substitutes his own ad-hoc claims that depart in significant respects from the IPCC's analysis on the effects of global warming.

For example, citing the 2007 IPCC assessment, Lomborg wrote in the main text of *Cool It:* "In its 'standard' future scenario, the IPCC predicts that the global temperature in 2100 will have risen on average 4.7°F from the current range."[8] In his endnote to this statement, Lomborg cited two sources from the 2007 IPCC assessment;[9] however, only one of the sources—an IPCC chart—could be located.[10] The second source—which Lomborg identified as "fig. 10.3.1" (Figure 10.3.1)—could not be located.[11]

Furthermore, the IPCC chart that could be found does not refer to any "standard" future climate scenario (as Lomborg wrote) that projects a 4.7°F "average" increase in global temperatures by 2100. Instead, the IPCC chart provides an array of socioeconomic conditions with a range of related projected climate outcomes, collectively referred to as "SRES scenarios."[12] "SRES" is an acronym for a 2000 IPCC publication titled, *Special Report on Emissions Scenarios*, which describes four basic scenarios (six such scenarios have since been issued), each depicting "different demographic, social, economic, technological, and environmental developments" for the twenty-first century, and thus "the driving forces" behind projected greenhouse emissions and temperature increases. Since the IPCC remained agnostic about which of the SRES development scenarios will prevail throughout the twenty-first century, it did not project any single "standard" or "average" levels of greenhouse emissions or temperature increases. In this regard, the IPCC's *Special Report on Emissions Scenarios* reported: "All [SRES scenarios] are equally valid with no assigned probabilities of occurrence."[13] It also reported: "Future greenhouse gas (GHG) emissions are the product of very complex dynamic systems, determined by driving forces such as demographic development, socio-economic development, and technological change. Their future evolution is highly uncertain. Scenarios are alternative images of how the future might unfold and are an appropriate tool with which to analyse how driving forces may influence future emission outcomes and to assess the associated uncertainties. They assist in climate change analysis, including climate modeling and the assessment of impacts, adaptation, and mitigation. The possibility that any single emissions path will occur as described in scenarios is highly uncertain."[14] In short, there is no IPCC "standard" or "average" future scenario (as Lomborg wrote) for climate change and average global temperatures in the IPCC assessments, including in the 2007 IPCC assessment report.

Furthermore, the SRES scenarios are not obscure concepts in the 2007 IPCC assessment report. They imbue many major sections of the report, and are fundamental to a competent public rendering of the IPCC's assessments. In fact, the SRES scenarios are referred to more than twenty times in the eighteen-page IPCC "Summary for Policymakers" that Lomborg claimed listed a "standard" future climate scenario. And the chart that Lomborg cited (that is, the one that could be located) is itself a depiction of six SRES scenarios projecting six ranges of temperature projections by 2100,[15] none of which is identified as a "standard" scenario.

Why is this worth mentioning? For one thing, at their most basic level, the SRES development scenarios reflect a consensus that the more we reduce greenhouse emissions, the lower global temperatures will be. In fact, the IPCC chart to which we just referred, and that Lomborg inaccurately cited as depicting a "standard" projected increase in global temperatures, projected an increase in global temperatures of about 0.6°C (1°F) from 2000 to 2100 if greenhouse emissions can be held constant at 2000 levels throughout the twenty-first century.[16] At the same time, the worst-case SRES development scenario, which reflects a high-growth, fossil-fuel-intensive global economy for the twenty-first century, depicts a worst-case warming of 2.4–6.4°C (4.3–11.5°F).[17] In short, by essentially disregarding the development, emissions, and climate features of the SRES scenarios, Lomborg circumvents a key IPCC concept, the implications of which inconveniently undermine Lomborg's Theorem with respect to potentially catastrophic temperature increases and Lomborg's Corollary that we do not need to reduce greenhouse emissions to any significant extent.

The two endnotes to the 2007 IPCC assessment discussed above—one to an IPCC chart that referenced a nonexistent "standard" future climate scenario, and the other to a nonexistent "fig. 10.3.1"—are the fifth and sixth of Lomborg's references to the IPCC in *Cool It* chapter 2. Lomborg's seventh reference to the IPCC supported the following uncontroversial statement: "The IPCC finds that the [warming] trends we have seen over the twentieth century will continue, with temperatures increasing more over land, more in the winter, and especially in the high northern latitudes: Siberia, Canada, and the Arctic."[18] And Lomborg's eighth reference supported this statement: "In the wintertime, temperatures might increase 9°F in Siberia compared to perhaps 5°F in Africa." This statement is referenced to "fig. 10.3.6" (Figure 10.3.6) in the 2007 IPCC assessment report, though no such figure could be located.[19]

Lomborg used his ninth, tenth, and eleventh references to the 2007 IPCC report to support the following statements: "We will see a marked decrease in frost days almost everywhere in the middle and high latitudes, and this will lead to a comparable increase in the growing-season length";[20] "Models show that heat events we now see every twenty years will become more frequent. By the end of the [twenty-first] century, we will have such events happening every three years";[21] "But cold spells will decrease just as much as heat waves increase. In areas where there is one cold spell every three years, by the end of the century such spells will happen only once every twenty years."[22] Lomborg then added, without an accompanying endnote, "*This means fewer deaths from cold, something we rarely hear about*" (emphasis added).[23]

Lomborg thus began his analysis of a major tenet of Lomborg's Theorem — that the benefits of warmer temperatures overall will reduce human mortality and morbidity, and that environmentalists cover up this hidden benefit of global warming. To support this last assertion, Lomborg wrote: "In the U.S. 2005 Climate Change and Human Health Impacts report, heat is mentioned fifty-four times and cold just once. It might seem callous to weigh lives saved versus those lost, but if our goal is to improve the lot of humanity, then it's important to know just how many more heat deaths we can expect compared to how many fewer cold deaths."[24] Lomborg then argued, "It seems reasonable to conclude from the data that, within reasonable limits, global warming might actually result in lower death rates."[25]

In reaching this conclusion, however, Lomborg boiled down the impact of global warming to its *direct* temperature-related impact on human mortality; but there are numerous *indirect* effects that Lomborg ignored upon asserting that global warming will reduce human death rates. Lomborg arrived at this conclusion after four hundred words of analysis. And the five sources that he cited within the four-hundred words failed to provide the "data" that Lomborg supposedly generated to support his claim that "global warming might actually result in lower death rates."[26]

Lomborg also ignored important sections in the 2007 IPCC assessment report that addressed the impact of global warming on human health, including a detailed and extensively documented chapter titled "Human Health."[27] This chapter looked at the projected *direct* impacts of global warming on human morbidity and mortality in several sections, all ignored by Lomborg: "Heat and Cold Health Effects" (Section 8.2.1); "Heatwaves" (Section 8.2.1.1); "Cold-Waves" (Section 8.2.1.2); "The European Heatwave 2003: Impacts and Adaptation" (Box 8.1); "Estimates of Heat and Cold Effects" (Section 8.2.1.3); "Heat-and Cold-Related Mortality" (Section 8.4.1.3), and "Projected Impacts of Climate Change on Heat-and Cold-Related Mortality" (Table 8.3).

Lomborg also ignored sections in the "Human Health" chapter that examined the *indirect* impacts of global warming on human morbidity and mortality: "Wind, Storms and Floods" (Section 8.2.2); "Drought, Nutrition and Food Security" (Section 8.2.3); "Food Safety" (Section 8.2.4); "Water and Disease" (Section 8.2.5); "Air Quality and Disease" (Section 8.2.6); "Ground-Level Ozone" (Section 8.2.6.1); "Effects of Weather on Concentrations of Other Pollutants" (Section 8.2.6.2); "Air Pollutants from Forest Fires" (Section 8.2.6.3); "Long-Range Transport of Pollutants" (Section 8.2.6.4); "Aeroallergens and Disease" (Section 8.2.7); "Vector-Borne, Rodent-Borne and Other Infectious Diseases" (Section 8.2.8); "Dengue" (8.2.8.1); "Climate Change, Migratory Birds and Infectious Disease" (Box. 8.4); "Malaria" (8.2.8.2); "Other Infectious Diseases" (Section 8.2.8.3); "Occupational Health (Section 8.2.9); "Ultraviolet Radiation and Health" (Section 8.2.10); "Projections of Climate-Change-Related Health Impacts" (Section 8.4.1); "Global Burden of Disease Study" (Section 8.4.1.1); "Malaria, Dengue and Other Infectious Diseases" (Section 8.4.1.2); "Urban Air Quality" (Section 8.4.1.4); "Projected Impacts of Climate Change on Malaria, Dengue Fever and Other Infectious Diseases" (Table 8.2), and "Projected Trends in Climate-Change-Related Exposures of Importance to Human Health" (Box 8.5).[28]

While Lomborg focused on the direct effects of hot and cold weather while concluding that global warming will lead to a net decrease in temperature-related human deaths, it is clear that the broader context—which includes the overall impacts of a warmer world on key factors such as food production, the availability of fresh water, and the spread of infectious disease, among others—should have been considered. In fact, the first paragraph of the executive summary of the "Human Health" chapter in the 2007 IPCC assessment reflects this broader approach: "Climate change currently contributes to the global burden of disease and premature deaths (very high confidence). Human beings are exposed to climate change through changing weather patterns (temperature, precipitation, sea-level rise and more frequent extreme events) and indirectly through changes in water, air and food quality and changes in ecosystems, agriculture, industry and settlements and the economy. At this early stage, the effects are small but are projected to progressively increase in all countries and regions."[29]

Furthermore, and after extensive analysis, the IPCC chapter on "Human Health" reached the opposite conclusion upon weighing the projected direct *and* indirect impacts of global warming on human health:

> Projected trends in climate-change-related exposures of importance to human health will:
>
> - Increase malnutrition and consequent disorders, including those relating to child growth and development (high confidence);
> - Increase the number of people suffering from death, disease and injury from heatwaves, floods, storms, fires and droughts (high confidence);
> - Continue to change the range of some infectious disease vectors (high confidence);
> - Have mixed effects on malaria; in some places the geographical range will contract, elsewhere the geographical range will expand and the transmission season may be changed (very high confidence);
> - Increase the burden of diarrhoeal diseases (medium confidence);
> - Increase cardio-respiratory morbidity and mortality associated with ground-level ozone (high confidence);
> - Increase the number of people at risk of dengue (low confidence);
> - *Bring some health benefits, including fewer deaths from cold, although it is expected that these will be outweighed by the negative effects of rising temperatures worldwide, especially in developing countries (high confidence).*[30] [Emphasis added]

In addition, as I indicated, though Lomborg cited five studies while arguing that "global warming might actually result in lower death rates" due to a decrease in cold-related deaths, none of the studies supported that claim.[31] For example, Lomborg cited one of the studies (Ebi et al., 2006) to complain that "heat is mentioned fifty-four times and cold just once."[32] Lomborg also cited two other sources (Basu and Samet, 2002; McMichael, Woodruff, and Hales, 2006) to complain that they "talk *only* about heat-related deaths."[33] Lomborg wrote nothing more about these important studies, the content of which are at odds with his claim that global warming will reduce human mortality.

The first of these studies—"Climate Change and Human Health Impacts in the United States: An Update on the Results of the U.S. National Assessment" (the "Ebi Report")[34]—is a review of the health-sector component of the *U.S. National Assessment of the Potential Consequences of Climate Variability and Change*, issued by the United States Global Change Research Program.[35] The *U.S. National Assessment* is a major U.S. government sponsored project on the impact of global warming. The release and distribution of the first *U.S. National Assessment* in 2000 was challenged by a conservative political organization, reportedly funded by the U.S. energy industry, as the *New York Times*'s Andrew Revkin reported: "An antiregulatory group sued the Bush administration yesterday to force it to stop distributing a report on climate change that the group contends is inaccurate and biased. The lawsuit was filed in Federal District Court in Washington [D.C.] by the Competitive Enterprise Institute, a group with industry support that contends global warming poses no significant risks."[36] Though Lomborg mentioned the Ebi Report in *Cool It*, he also discarded its findings.

In light of Lomborg's summary dismissal of the Ebi Report and, by extension, the "Health Sector Assessment" (HSA) of the *U.S. National Assessment*, it is worth noting that the Ebi Report "focused on five categories of health outcomes: temperature-related morbidity and mortality, the health impacts of extreme weather events (e.g., storms and floods), health outcomes associated with air pollution, water-and food-borne diseases, and vector-and rodent-borne diseases." The Ebi Report continued: "The integrated assessment approach that was used [by the HSA] reviewed a wide-range of literature on climate and health, relied on the expert judgment of the health sector team and those with whom they consulted, and incorporated, where available, some limited modeling of the projected impacts of climate on health." In summary, the Ebi Report noted: "The Health Sector Assessment concluded that climate variability and change are likely to increase morbidity and mortality risks for several climate-sensitive health outcomes, with the net impact uncertain."[37] While arguing in *Cool It* about how global warming would likely lead to a net reduction in human mortality due to fewer cold-related deaths, and though he cited the Ebi Report (to complain that it mentioned heat more than cold), Lomborg neglected to mention this clearly relevant portion of the report.

The second study that Lomborg also quickly cited and dismissed (Basu and Samet, 2002) focused primarily on the direct effects of temperature-related mortality, and emphasized the potential threat as well as the current difficulty of discerning the future health impact of global warming on human mortality: "The effect of elevated temperature on mortality is a public health threat of considerable magnitude. . . . Models of the relation between temperature and mortality are needed to predict the consequences of global warming, particularly for those most vulnerable and least able to adapt."[38]

The third of the three studies (McMichael, Woodruff, and Hales, 2006) concluded: "Climate change will affect human health in many ways—mostly adversely," though the authors also noted "the residual uncertainties in modelling" with respect to the health impact.[39]

Given that the largest of these Lomborg-cited studies reported that global warming is "likely to increase morbidity and mortality risks for several climate-sensitive health outcomes," and that all three studies nevertheless reported some uncertainty about the future net impact of global warming on human health, it is clear that these studies do not support Lomborg's assertion that "it seems reasonable to conclude from the data that, within reasonable limits, global warming might actually result in lower death rates." This leaves only two studies from which Lomborg could have derived the "data" that led to this conclusion.

Lomborg cited the first of the two remaining studies (Martens,

1998) to support the following statement in *Cool It:* "For almost every location in the world, there is an 'optimal' temperature at which deaths are the lowest. On either side of this temperature – both when it gets colder and warmer – death rates increase."[40] Lomborg's suggestion to the contrary, this statement by itself does not support Lomborg's claim that human mortality will decrease due to global warming. Lomborg then used the remaining study (Keatinge et al., 2000) to support the next statement: "However, what the optimal temperature is is a different issue. If you live in Helsinki, your optimal temperature is about 59°F, whereas in Athens you do best at 75°F."[41] Likewise, this statement does not support Lomborg's contention that global warming will be likely to reduce human mortality. Thus, the five Lomborg-cited studies do not provide the "data" that Lomborg claimed supported his conclusion that "global warming might actually result in lower death rates."

This ended the referenced portion in the first section of Lomborg's claim that global warming will lead to a net reduction in human mortality due to fewer cold-related deaths. What followed in *Cool It* was a detailed but unreferenced continuation of the same claim:

> The important point to notice is that the best temperature is typically very similar to the average summer temperature. Thus, the actual temperature will only rarely go above the optimal temperature, but very often it will be below. In Helsinki, the optimal temperature is typically exceeded only 18 days per year, whereas it is below that temperature a full 312 days. Research shows that although 298 extra people die each year from it being too hot in Helsinki, some 1,655 people die from it being too cold.
>
> It may not be so surprising that cold kills in Finland, but the same holds true in Athens. Even though absolute temperatures of course are much higher in Athens than in Helsinki, temperatures still run higher than the optimum one only 63 days per year, whereas 251 days are below it. Again, the death toll from excess heat in Athens is 1,376 people each year, whereas the death toll from excess cold is 7,852.
>
> This trail of statistics leads us to two conclusions. First, we are very adaptable creatures. We live well both at 59°F and 75°F. We can adapt

> to both cold and heat. Further adaptation on account of global warming will not be unproblematic, because we have already invested heavily in housing and infrastructure such as heating and air-conditioning to handle our current climate. But that is why the second point is so important. It seems reasonable to conclude from the data that, within reasonable limits, global warming might actually result in lower death rates.[42]

Given that this conclusion is a major plank of Lomborg's Theorem—that "within reasonable limits, global warming might actually result in lower death rates"—it is surprising that not one factual assertion or statistic in the above text appears to be attributed to any source. Even when we attempt to fill in the blanks ourselves, and assume that Lomborg derived these statistics from one or more of the five sources that we just reviewed, Lomborg's statistics, as quoted above, still cannot be found, though Keatinge et al. reported roughly comparable figures. Thus, when Lomborg wrote that "research shows that although 298 extra people die each year from it being too hot in Helsinki, some 1,655 people die from it being too cold," the statistics reported in Keatinge are that "annual heat related mortality" in Helsinki ("South Finland") is 248, and "annual cold related mortality" is 1,379. Likewise, though Lomborg wrote that "the death toll from excess heat in Athens is 1,376 people each year, whereas the death toll from excess cold is 7,852," Keatinge reported that "annual heat related mortality" in Athens is 445, and "annual cold related mortality" is 2,533.[43] Beyond Keatinge, it is unclear where Lomborg's numbers came from. Furthermore, as its abstract indicates, the subjects of the Keatinge study were "people aged 65–74," not the undifferentiated category of "people" used by Lomborg.

Regardless of the origin of Lomborg's statistics—even assuming that they were accurately quoted from a legitimate source—they do not substantiate his conclusion that fewer cold-related deaths will more than offset the additional heat-related deaths worldwide. For example, the statistics just quoted involve two relatively affluent European cities (Helsinki and Athens); however, the IPCC assessment chapter titled "Human Health" reported: "Adverse health impacts will be greatest in low-income countries."[44]

Lomborg also addressed the 2003 heat wave in Europe, alleging that Al Gore had misconstrued its significance. Lomborg quoted Gore as saying, "We have already begun to see the kind of heat waves that scientists say will become much more common if global warming is not addressed. In the summer of 2003 Europe was hit by a massive heat wave that killed 35,000 people."[45] Lomborg responded: "Yet while we will see more and hotter heat waves, talking only about heat waves means we leave out something even more important," referring to his contention that a reduction in cold-related deaths will outweigh heat-related deaths.[46] A few pages later, Lomborg wrote that a fall 2003 report by the Earth Policy Institute[47] helped fuel "the public perception that the [2003] heat wave became a sure indicator of global warming." Lomborg objected: "But group wisdom can occasionally be wrong."[48]

To support his view that Al Gore and the Earth Policy Institute were wrong about linking the 2003 heat wave to global warming, Lomborg cited a 2006 study published in *Geophysical Research Letters:*[49] "A recent academic paper has checked this theory" that the 2003 heat wave was due to global warming and that more such heat waves would occur. Lomborg observed that the paper "concluded that although the circumstances [of the 2003 heat wave] were unusual, equal or more unusual warm anomalies have occurred since 1979."[50] Lomborg's point here is unclear, since the period 1979–2003 is recent and short, and within which the effects of global warming were already becoming apparent. In addition, while Lomborg cited this 2006 study, he neglected to cite a 2005 study published in *Nature,* which was very much at odds with Lomborg's claim that the 2003 heat wave was neither unusual nor attributable to global warming:

> The summer of 2003 was probably Europe's hottest summer in over 500 years, with average temperatures 3.5°C [6.3°F] above normal. With approximately 22,000 to 45,000 heat-related deaths occurring

> across Europe over two weeks in August 2003, this is the most striking recent example of health risks directly resulting from temperature change. . . . The European heatwave in 2003 was well outside the range of expected climate variability. In addition, comparisons of climate model outputs with and without anthropogenic drivers show that the risk of a heatwave of that magnitude had more than doubled by 2003 as a result of human-induced climate change. The demonstration of a causal link between global warming and the occurrence of regional heatwaves indicates a potential for more frequent and/or more severe heatwaves in a future warmer world.[51]

Though he later cited the *Nature* report, Lomborg did not quote or otherwise allude to these words.[52]

Lomborg also did not mention that the 2001 IPCC assessment report projected that "more hot days and heat waves over nearly all land areas" would occur this century because of global warming.[53] The IPCC listed the confidence level of this projection as "very likely," which reflects a greater than 90 percent likelihood of accuracy.[54]

Lomborg also ignored what the 2007 IPCC assessment report said about the 2003 European heat wave: "The TAR [the 2001 IPCC assessment report] noted a very likely increase in the intensity and frequency of summer heatwaves throughout Europe, and one such major heatwave [in 2003 in Europe] has occurred since the TAR."[55] The 2007 IPCC Assessment also reported that "European summer climate would experience a pronounced increase in year-to-year variability and thus a higher incidence of heatwaves and droughts."[56] And it reported: "Over the next century, heatwaves are very likely to become more common and severe."[57] Thus, while he argued that Al Gore and the Earth Policy Institute had inappropriately linked the 2003 heat wave in Europe to global warming, Lomborg ignored the relevant portions of a 2005 study published in *Nature*, in addition to the 2001 and 2007 IPCC assessment reports, all of which expressed confidence in a connection between global warming and a higher incidence of heat waves in the recent past and projected future.

A few paragraphs later Lomborg wrote, again problematically: "In Europe as a whole, about two hundred thousand people die from excess heat each year. However, about 1.5 million Europeans die annually from excess cold. That is more than seven times the total number of heat deaths."[58] But Lomborg's only referenced source for these figures—a chart in the statistical annex of a 2004 World Health Organization report—contains no data on human mortality due to excess heat or cold.[59] In fact, the words "excess heat" and "excess cold" make no appearance in the WHO document; neither does the word "heat," and the word "cold" appears only once in a reference unrelated to death due to excess cold.

Lomborg's reference to the WHO document, which allegedly supports his claim that two hundred thousand people die each year in Europe from excess heat, reads in its entirety: "207,000, based on a simple average of the available cold and heat deaths per million, cautiously excluding London and using WHO's estimate for Europe's population of 878 million (WHO, 2004a:121)."[60] However, page 121 of the 2004 WHO report—*The World Health Report 2004: Changing History*—which is what this source references, lists no data on cold- and heat-related deaths per million, or for cold- and heat-related deaths in any context.[61] Likewise, Lomborg's very next reference—to support his claim that 1.5 million Europeans die annually from excess cold—reads in its entirety: "1.48 million, estimated in the same way as total heat deaths."[62] Thus, Lomborg's references indicate that page 121 of the 2004 WHO report is the source of his estimates of annual heat- *and* cold-related deaths in Europe; however, this page in the WHO report lists no statistics for either cold- or heat-related deaths. Consequently, there is no apparent basis for Lomborg's claim that 1.5 million Europeans die annually from excess cold.

Despite the apparent absence of a factual basis for Lomborg's key assertion that 1.5 million Europeans die each year from excess cold, Lomborg took this unsubstantiated estimate of annual cold-related deaths in Europe, multiplied it by ten, and wrote: "Just in the past decade, Europe has lost about fifteen million people to the cold, more than four hundred times the iconic heat deaths from 2003. That we so easily neglect these deaths and so easily embrace those caused by global warming tells us of a breakdown in our sense of proportion."[63]

This passage is not referenced. Because Lomborg presents no actual source for his claim that 1.5 million Europeans die annually from cold-related causes, and because that estimate is the predicate for Lomborg's claim that 15 million people died in Europe from excess cold in some ten-year period that preceded the publication of *Cool It*, it seems that Lomborg presented no evidence that either 1.5 million Europeans die annually, or that 15 million Europeans died because of cold weather in a recent decade.

Continuing, Lomborg asks: "How will heat and cold deaths change over the coming century?"[64] He answers first by referring to three studies on projected heat- and cold-related deaths *in Europe*;[65] however, studies on European populations do not address global mortality rates. Lomborg then cited a fourth study: "Indeed, a paper trying to incorporate all studies on this issue and apply them to a broad variety of settings both developed and developing around the world found that 'global warming may cause a decrease in mortality rates, especially of cardiovascular diseases.'"[66] But this Lomborg-cited paper was published in 1998,[67] and thus would have been superseded by the 2001 and 2007 IPCC assessment reports. And the 2007 assessment report, which updated the IPCC's position on the impact of global warming on human health, and which took into account several relevant studies, stated: "Overall, climate change is projected to have some health benefits, including reduced cold-related mortality, reductions in some pollutant-related mortality, and restricted distribution of diseases where temperatures or rainfall exceed upper thresholds for vectors or parasites. However, the balance of impacts will be overwhelmingly negative." And: "The analyses suggest that climate change will bring some health benefits, such as lower cold-related mortality and greater crop yields in temperate zones, but these benefits will be greatly outweighed by increased rates of other diseases, particularly infectious diseases and malnutrition in low-income countries."[68]

To conclude his analysis on global warming and human mortality rates, Lomborg wrote:

> The first complete survey for the world was published in 2006, and what it shows us very clearly is that climate change will not cause massive disruptions or huge death tolls. Actually, the direct impact of climate change in 2050 will mean *fewer* dead, and not by a small amount. In total, about 1.4 million people will be saved each year, due to more than 1.7 million fewer deaths from cardiovascular diseases and 365,000 more deaths from respiratory disorders. This holds true for the United States and Europe (each with about 175,000 saved), as for the rest of the industrialized world. But even China and India will see more than 720,000 saved each year, with deaths avoided outweighing extra deaths nine to one. The only region where deaths will outweigh lives saved is in the rest of the developing world, especially Africa. There almost 200,000 deaths will be avoided, but more than 250,000 will die.[69]

Although Lomborg refers to this study as "the first complete survey for the world," the authors of that survey (Bosello, Roson, and Tol, 2006) had a more modest view of their research, which in fact assessed the effects of global warming on only six types of disease (cardiovascular disease, respiratory disease, diarrhea, malaria, dengue fever, and schistosomiasis). Although acknowledging that four of these diseases are "major killers," the authors also clarified their reasons for focusing on these six diseases and excluding others: "For other diseases probably affected by climate change, no global estimates are available. Our selection of diseases is therefore one of convenience, rather than comprehensiveness."[70] The authors thus excluded several projected warming-related causes of human mortality, including many infectious diseases, malnutrition and starvation, stroke, and drought. These exclusions are major limitations on what Lomborg referred to as "the first complete survey for the world" of global warming and its health-related impacts. For example, while Bosello et al. omitted drought-related health impacts, the 2007 IPCC assessment report concluded that "drought-affected areas are projected to increase in extent, with the potential for adverse impacts on multiple sectors, e.g. agriculture, water supply, energy production and health."[71] Elsewhere, the 2007 IPCC assessment projected "increased morbidity and mortality from heat waves, floods and droughts."[72]

Lomborg also ignored the results of studies which concluded that global warming would *increase* cardiovascular disease and human mortality. For example, one study (Basu and Samet, 2002),[73] citing a 2000 U.S. National Research Council report,[74] stated: "Major predicted health effects of long-term climatic change include skin and eye damage from increased exposure to ultraviolet radiation, increased incidence of respiratory and cardiovascular diseases, increased incidence of vector-borne and water-borne diseases, and heat-related morbidity and mortality."[75] Citing six published studies,[76] Basu and Samet also reported: "In studies of heat waves and elevated temperature, cardiovascular diseases, respiratory diseases, and cerebrovascular diseases were prominent causes of death."[77] And Basu and Samet cited other studies showing increases in total mortality (including an increase in cardiovascular disease) related to high temperatures, hot days, and heat waves.[78]

Though a major plank of Lomborg's Theorem is that global warming will reduce human mortality due to an offsetting reduction in cold-related deaths, he provides little to no evidence in *Cool It* that supports this position.

five
ON MELTING GLACIERS AND RISING SEA LEVELS

Though Lomborg never demonstrated in chapter 2 of *Cool It* that global warming would lead to a net reduction in human mortality, he began chapter 3 by claiming to have settled that issue, and by pledging to expose "many other" environmentalists' "exaggerations" concerning the impacts of global warming: "In chapter 2, we looked at what happens just when temperatures increase and saw that it was no catastrophe. But, of course, there are many other concerns about global warming, each often presented as a disaster-in-waiting, urging us to drop everything else and focus on cutting CO_2. As it turns out, these statements are often grossly exaggerated and divert us from making sound policy judgments. Let's take a realistic look at some of them."[1]

The global warming "exaggerations" that Lomborg supposedly uncovered in chapter 3, which he titled "Global Warming: Our Many Worries," were listed in the chapter as follows: "Melting Glaciers," "Rising Sea Levels," "Penguins in Danger?" "Extreme Weather, Extreme Hype," "Flooding Rivers," "A New Ice Age over Europe," "Malaria in Vermont," "More Heat Means More Starvation," and "Water Shortages." The first two of these issues—melting glaciers and rising sea levels—will be examined in this chapter. Most others will be examined in subsequent chapters.

Melting Glaciers

Before examining what Lomborg wrote about melting glaciers, it would first be informative to review what the 2007 IPCC assessment report said about the issue. The IPCC's glossary defines a glacier as "a mass of land ice that flows downhill under gravity," which is maintained by an "accumulation of snow at high altitudes" and "balanced by melting at low altitudes or discharge into the sea."[2] Beyond defining what a glacier is, it's also important to note that glaciers are part of the Earth's *cryosphere*, defined by the IPCC as "the component of the climate system consisting of all snow, ice and frozen ground (including permafrost) on and beneath the surface of the Earth and ocean."[3] The 2007 IPCC assessment also described the general features of the Earth's cryosphere as follows:

> Currently, ice permanently covers 10% of the land surface, with only a tiny fraction occurring outside Antarctica and Greenland. Ice also covers approximately 7% of the oceans in the annual mean. In midwinter, snow covers approximately 49% of the land surface in the NH [Northern Hemisphere]. An important property of snow and ice is its high surface albedo [the fraction of solar radiation reflected by the Earth's surface, expressed as a percentage[4]]. Because up to 90% of the incident solar radiation is reflected by snow and ice surfaces, while only about 10% is reflected by the open ocean or forested lands, changes in snow and ice cover are important feedback mechanisms in climate change. . . . The cryosphere stores about 75% of the world's freshwater. At a regional scale, variations in mountain snowpack, glaciers and small ice caps play a crucial role in freshwater availability.[5]

Upon describing the cryosphere, which includes glaciers (land ice), the 2007 IPCC assessment described the effects of global warming on the Earth's snow, ice, and frozen ground: "Since the change from ice to liquid water occurs at specific temperatures, ice is a component of the climate system that could be subject to abrupt change following sufficient warming. Observations and analyses of changes in ice have expanded and improved since the TAR [the 2001 IPCC assessment report], including shrinkage of mountain glacier volume, decreases in snow cover, changes in permafrost and frozen ground, reductions in Arctic sea ice extent, coastal thinning of the Greenland Ice Sheet exceeding inland thickening from increased snowfall, and reductions in seasonally frozen ground and river and lake ice cover."[6]

The 2007 IPCC assessment then itemized its findings in a more specific fashion with respect to the impact of global warming on the Earth's cryosphere. These included:

- Snow cover has decreased in most regions [of the world], especially in spring;
- The maximum area covered by seasonally frozen ground decreased by about 7% in the NH [Northern Hemisphere] over the later half of the 20th century, with a decrease in spring up to 15%;
- Annual average Arctic sea ice extent has shrunk by about 2.7 percent per decade since 1978 based upon satellite observations;
- During the 20th century, glaciers and ice caps have experienced widespread mass losses and have contributed to sea level rise.[7]

Given that glaciers are part of the Earth's cryosphere, and that most of the Earth's cryosphere is melting or thawing due to global warming, it would have made sense for Lomborg to situate the issue of melting glaciers within the context of the cryosphere, as the 2007 IPCC assessment report did. However, while arguing that the issue of "melting glaciers" is exaggerated, Lomborg never mentioned that ice and snow generally are melting and breaking up on a global scale. Nor does the word "cryosphere" appear in *Cool It.* This permitted Lomborg to focus on the issue of melting glaciers without situating that phenomenon within the larger context of the melting and thawing cryosphere, which is a powerful indicator of the effects of global warming.

In addition to the 2007 IPCC report's conclusions about global warming and the Earth's cryosphere, the 2001 IPCC assessment report—published six years before *Cool It*—also linked the melting of the Earth's cryosphere (including glaciers) to global warming:

- Changes in sea level, snow cover, ice extent, and precipitation are consistent with a warming climate near the Earth's surface. Examples of these include . . . widespread retreat of non-polar glaciers . . . and decreases in snow cover and sea-ice extent and thickness;[8]
- Decreasing snow cover and land-ice extent continue to be positively correlated with increasing land-surface temperatures. . . . There is now ample evidence to support a major retreat of alpine and continental glaciers in response to 20th century warming;[9]
- Glaciers and ice caps will continue their widespread retreat during the 21st century and Northern Hemisphere snow cover and sea ice are projected to decrease further. . . . Modeling studies suggest that

> the evolution of glacial mass is controlled principally by temperature changes, rather than precipitation changes, on the global average.[10]

Given the 2007 and 2001 IPCC assessments on the impact of global warming on the cryosphere, including glaciers, it is clear that Lomborg had to attempt to overcome overwhelming evidence to show that scientists and environmentalists had exaggerated the effects of global warming on glaciers.

Furthermore, upon updating its assessment of the impact of global warming on the cryosphere, the 2007 IPCC assessment included a new chapter titled, "Observations: Changes in Snow, Ice and Frozen Ground."[11] Lomborg completely ignored this chapter in his assessment of "Melting Glaciers" in *Cool It*. With respect to global warming and glaciers, the new IPCC chapter reported: "Glaciers and ice caps provide among the most visible indications of the effects of climate change."[12]

In addition—and this point is highly relevant to Lomborg's analysis—the 2007 IPCC assessment report noted that the atmospheric warming responsible for these changes in the Earth's cryosphere, including a significant melting of glaciers, was not due to natural causes, that is, increased solar insolation (the amount of solar radiation reaching the Earth): "The present day near-global retreat of mountain glaciers cannot be attributed to the same natural causes [as the retreat of glaciers thousands of years ago], because the decrease of summer insolation during the past few millennia in the Northern Hemisphere should be favourable to the growth of the glaciers."[13]

And equally relevant to Lomborg's analysis, as we shall see, the 2007 IPCC assessment also reported: "It is *very likely* that average Northern Hemisphere temperatures during the second half of the 20th century were higher than for any other 50-year period in the last 500 years. It is also *likely* that this 50-year period was the warmest Northern Hemisphere period in the last 1.3 kyr [1,300 years], and that this warmth was more widespread than during any other 50-year period in the last 1.3 kyr. . . . The rise in surface temperatures since 1950 *very likely* cannot be reproduced without including anthropogenic greenhouse gases in the model forcings, and it is *very unlikely* that this warming was merely a recovery from a pre-20th century cold period"[14] (emphasis in original). Thus, what the IPCC reported with near certainty—in 2001 and 2007—was that human-induced global warming is melting the Earth's cryosphere, including glaciers, on a global scale, and that the worldwide melting and thawing cannot be attributed solely to natural climate cycles or warming from the sun.

In his section on "Melting Glaciers," and while immediately tacking away from the IPCC's explanation about why the Earth's glaciers are melting, Lomborg began by describing the Medieval Warm Period and the Little Ice Age in order to establish a context for his explanation that the Earth's glaciers are melting due to a natural thawing from the Little Ice Age. Lomborg noted that "over the past millennium, temperatures have gone up and down and up again from natural causes." Lomborg devoted three pages of his seven-page section on melting glaciers to brief accounts of the climatic effects of the Medieval Warm Period and the Little Ice Age, including the observation that "the warmer climates and reduced sea ice made possible the colonization of the otherwise inhospitable Greenland and Vinland (Newfoundland) by the Vikings" during the Medieval Warm Period. Consistent with his thesis that global warming is "no catastrophe," Lomborg added: "It is perhaps worth noticing that while many events during the Little Ice Age are seen and reported as negative, this does not seem to be the case with most of the Medieval Warm Period."[15]

Though the 2007 IPCC assessment concluded that it is "very unlikely" that the current warming is "merely a recovery from a pre-20th century cold period," including the so-called Little Ice Age, this explanation constitutes at least half of Lomborg's account for today's melting glaciers, though he also noted that "in the past 150 years, temperatures have diverged even more upward due to global warming." He thus observed that while "it is clear that part of the temperature increase since [the Little Ice Age] has simply been a result of coming out of the Little Ice Age," it "is also clear, though, that we are now seeing a warming trend beyond that, indicating man-made global warming." Lomborg concludes that "both of these warmings" — that is, the natural emergence from the Little Ice Age and the recent human-induced warming — "have caused glaciers to recede."[16]

But in the very next sentence (after conceding that at least part of the glacial melting is the result of human-induced global warming), Lomborg wrote: "Many have seized on pictures of these retreating glaciers as symbols of global warming. Al Gore, for example, fills eighteen pages of his book with before-and-after pictures of glaciers."[17] It seems that Lomborg's complaint here is that Gore did not acknowledge that the glaciers are melting due to the Earth's emergence from the Little Ice Age. However, as Lomborg must or should have known, the 2007 IPCC assessment report stated that it is "very unlikely" that the Earth is warming due to "a recovery from a pre-20th century cold period."

Furthermore, the 2001 IPCC assessment report also reported that the Medieval Warm Period and the Little Ice Age were likely regional and not global climate events:

> The terms "Little Ice Age" and "Medieval Warm Period" have been used to describe two past climate epochs in Europe and neighbouring regions during roughly the 17th to 19th and 11th to 14th centuries, respectively. The timing, however, of these cold and warm periods has recently been demonstrated to vary geographically over the globe in a considerable way. Evidence from mountain glaciers does suggest increased glaciation in a number of widely spread regions outside Europe prior to the 20th century, including Alaska, New Zealand and Patagonia. However, the timing of maximum glacial advances in these regions differs considerably, suggesting that they may represent largely independent regional climate changes, not a globally-synchronous increased glaciation.[18]

The 2001 IPCC assessment report stated further that "the 'Little Ice Age' appears to have been most clearly expressed in the North Atlantic region as altered patterns of atmospheric circulation" and that "Medieval warmth appears, in large part, to have been restricted to areas in and neighbouring the North Atlantic."[19] The IPCC also reported that evidence of temperature changes in past centuries in the

Southern Hemisphere "is quite sparse," though suggesting "markedly different behavior from the Northern Hemisphere" with "the only obvious similarity" between the two hemispheres being "the unprecedented warmth of the late 20th century."[20] Likewise, the 2007 IPCC assessment concluded: "The evidence currently available indicates that NH [Northern Hemisphere] mean temperatures during medieval times (950–1100) were indeed warm in a 2-kyr [two thousand year] context and even warmer in relation to the less sparse but still limited evidence of widespread average cool conditions in the 17th century. However, the evidence is not sufficient to support a conclusion that hemispheric mean temperatures were as warm, or the extent of warm regions as expansive, as those in the 20th century as a whole, during any period in medieval times."[21]

It thus seems that Lomborg's effort to situate today's worldwide glacial melting within the context of a natural emergence from the Little Ice Age is further weakened by evidence that these warm and cold periods were most likely phenomena specific to the North Atlantic region. Lomborg's explanation is also weakened by IPCC-reported evidence that the warming in medieval times was overall neither as warm nor as global as today's warming.

After seeking to tie today's melting glaciers to natural climate variations over the past thousand years, Lomborg then sought to situate the glacial melting in the context of the past ten thousand years, writing that "several facts impede [Gore's] rather simple narrative" of melting glaciers due to human-induced global warming.[22] "First," Lomborg wrote, "glaciers have been greatly advancing and receding since the last ice age." A few sentences later, in one of only two references to an IPCC report in his section on melting glaciers, Lomborg wrote: "In fact, most glaciers in the Northern Hemisphere were small or absent from nine thousand to six thousand years ago."[23] Lomborg referenced this statement to a sidebar in a chapter in the 2007 IPCC assessment on "Paleoclimate," which, as Lomborg noted, reported as follows: "Most archives from the NH [Northern Hemisphere] and the tropics indicate short, or in places even absent, glaciers between 11 and 5 ka, whereas during the second half of the Holocene, glaciers reformed and expanded. This tendency is most probably related to changes in summer insolation due to the configuration of orbital forcing."[24]

Could it be, given Lomborg's argument and by citing the IPCC to this effect, that today's worldwide glacial melting can be explained by changes in "solar insolation" and "orbital forcing"—natural phenomena upon which humans have no influence? To answer this question, and by way of explanation, Lomborg used the top half of the sidebar in question to advance his claim that glaciers have come and gone in the past due to such natural influences, but he ignored the bottom half, which acknowledged "gaps" in the data but nevertheless pointed out that such swings in melting and returning glaciers took place over periods that were thousands of years long, that there is no record or evidence of such swings taking place on a global scale within a thousand-year period, and that the current, rapid rate of glacial melting cannot be explained by long-term changes in solar insolation. Here is what the same IPCC sidebar reported in this regard:

> Comparing [today's] ongoing retreat of glaciers with the reconstruction of glacier variations during the Holocene [which dates back to about eleven thousand years], no period analogous to the present with a globally homogenous trend of retreating glaciers over centennial [a thousand years] and shorter time scales could be identified in the past, although account must be taken of the large gaps in the data coverage on retreated glaciers in most regions. This is in line with model experiments suggesting that present-day glacier retreat exceed any variations simulated by the GCM [general circulation models] control experiments and must have an external cause, with anthropogenic forcing [human-caused changes] the most likely candidate.[25]

This statement from the IPCC sidebar—that no period within the Holocene with respect to expanding and receding glaciers is analogous to today's melting glaciers—undermines an entire page (page 55) in Lomborg's *Cool It*, wherein he meaningfully describes glaciers receding and expanding throughout the Holocene.

Similarly, and on the same page (page 55), Lomborg wrote: "While glaciers since the last ice age have waxed and waned, they overall seem to have been growing bigger and bigger each time until reaching their absolute maximum at the end of the Little Ice Age." Lomborg referenced this statement to a 2000 study published in the *Annals of Glaciology*, which examined the retreat and expansion of glaciers in Norway, Switzerland, and New Zealand. Lomborg's point was revealed in the following unreferenced comment in *Cool It:* "So it is not surprising that as we're leaving the Little Ice Age we are seeing glaciers dwindling. We are comparing them with their absolute maximum over the past ten millennia."[26] This contributed to Lomborg's argument that today's glacial melting is due to the natural "waxing and waning" of glaciers. And while Lomborg argued along these lines, he neglected to mention the conclusion of the same 2000 study in *Annals of Glaciology:* "The current retreat [of glaciers] cannot be explained from natural variability in glacial length and must be due to external [human-induced] forcing."[27]

Lomborg also linked the melting of glaciers on Mount Kilimanjaro in Tanzania, Africa, to his claim that glaciers are melting worldwide as a natural recession from their peak in the Little Ice Age: "So it is not surprising that as we're leaving the Little Ice Age we are seeing glaciers dwindling. We are comparing them with their absolute maximum over the past ten thousand millennia. The best-documented overview of glaciers shows that they have been receding continuously since 1800. The perfect glacier icon, the snow-clad Mount Kilimanjaro, has been receding at least since 1880. When Ernest Hemingway published 'The Snows of Kilimanjaro' in 1936, the mountain had already lost more than half its glacier surface area in the previous half century."[28]

Lomborg thus presents the glacial melting on Kilimanjaro as a high-profile illustration of his notion that glaciers worldwide are receding naturally from their peak during the Little Ice Age. But among the rest of the world's glaciers, the causes of melting glaciers on Kilimanjaro, at least as currently studied, are an exception to the causes of glacier-melting worldwide. Citing a study by several scientists, Lomborg wrote that "Kilimanjaro has not lost its ice on account of increasing temperatures, which have remained rather stable below freezing, but because of a regional shift around 1880 toward drier climates."[29] The study cited by Lomborg reported that "a drastic drop in atmospheric moisture at the end of the 19th century and the ensuing drier climatic conditions are likely forcing glacier retreat on Kilimanjaro."[30] The 2007 IPCC assessment report concurred with these findings, noting that the "glaciers on Kilimanjaro behaved *exceptionally* throughout the 20th century."[31] This means that the hypothesis that Kilimanjaro's glaciers are shrinking due to decreased precipitation unrelated to global warming, if true, would constitute an exception to the evidence that glaciers are melting worldwide because of global warming and higher temperatures.[32]

Lomborg himself acknowledges this point; nevertheless, and immediately following this acknowledgement, he inexplicably applies the exceptional nature of Kilimanjaro's melting glaciers to glaciers worldwide, then concludes in non-sequitur fashion that reducing greenhouse emissions will not save the glaciers:

> Furthermore, Kilimanjaro has not lost its ice on account of increasing temperatures, which have remained rather stable below freezing, but because of a regional shift around 1880 toward drier climates. Thus, Kilimanjaro is not a good poster child for man-made global warming. In the latest satellite study, it is concluded that "results suggest glaciers on Kilimanjaro are merely remnants of a past climate rather than sensitive indicators of 20th century climate change."
>
> Yet we are often told that we need to reduce CO_2 emissions to address the problem of the receding glaciers. In a video with Kilimanjaro in the background, Greenpeace tells us that the mountain's entire ice field might be lost by 2015 due to climate change: "This is the price we pay if climate change is allowed to go unchecked." But of course, for Kilimanjaro we are able to do nothing, since it is losing ice due to a drier climate. Even if we granted that its demise was partially related to global warming, nothing we could do would have even the slightest impact before 2015.[33]

Given the above context, it seems that Lomborg's complaint—"Yet we are often told that we need to reduce CO_2 emissions to address the problem of the receding glaciers"—was slipped into his discussion of Kilimanjaro like a cardplayer cheating at poker. While reducing greenhouse emissions may do little to prevent the loss of glaciers on Kilimanjaro, reducing such emissions could reduce the melting of other glaciers worldwide, since greenhouse emissions represent the prevailing cause of the melting.

One page later, Lomborg referred again to Kilimanjaro with an off-topic reference to Tanzanian farmers: "While emotionally charged pictures of the beautiful glaciers from Kilimanjaro paired with admonishing concerns over CO_2 undoubtedly are very effective with the media and opinion makers, they hardly address the real problems of the Tanzanian farmers on the slopes." While any impact of global warming upon "the Tanzanian farmers on the slopes" of Kilimanjaro would be a valid point in a book about global warming, the manner in which Lomborg introduces the issue here is disingenuous, since the glaciers of Kilimanjaro, and the farmers on its slopes, appear to be unrelated to global warming. Nevertheless, Lomborg wrote: "I believe we have to dare to ask whether we help Tanzanians best by cutting CO_2, which would make no difference to the glaciers, or through HIV policies that would be cheaper, faster, and have much greater effect."[34] This line of reasoning is an excellent example of Lomborg's *modus operandi:* an inaccurate assumption (that glaciers are melting worldwide due primarily to a natural thaw from the Little Ice Age), surrounded by invalid inferences (pertaining to the emissions-reduction implications for Kilimanjaro), wrapped with a non sequitur (the invocation of the Tanzanian farmers on the slopes of Kilimanjaro), leading to an erroneous conclusion (that reducing greenhouse emissions is bad policy).

Lomborg then changed the focus from Mount Kilimanjaro to the glaciers on the Himalayan Mountains. Though the 2007 IPCC assessment report provided a significant amount of analysis of the impact of global warming on Himalayan glaciers,[35] Lomborg ignored it (with one exception below), and was thus unconstrained in pursuing the Little Ice Age connection to the Himalayan glaciers, while also misstating the projected life expectancy of those glaciers. Thus, Lomborg wrote in *Cool It:* "Glaciers in the Himalayas have been declining significantly since the end of the Little Ice Age and have caused increasing water availability throughout the centuries, possibly contributing to higher agricultural productivity. But with continuous melting, the glaciers will run dry toward the end of the century."[36]

Lomborg referenced his assertion that the Himalayan glaciers have been declining "since the end of the Little Age" to a 2005 study in the science journal *Boreas,* which reported: "Since the Little Ice Age, and particularly during this century, glaciers [in the Himalayas] have been progressively retreating. This pattern is likely to continue throughout the 21st century, exacerbated by human-induced global warming."[37] Lomborg invoked only the part that referenced the Little Ice Age, and chopped off the part that invoked "human-induced global warming."

Lomborg's assertion that the Himalayan glaciers will run dry "toward the end of the [twenty-first] century" was referenced to the 2007 IPCC assessment report (Working Group II, Chapter 3, Section 3.4.1). Although Section 3.4.1 spans four pages (pp. 182–185) and includes two charts (Figures 3.3 and 3.4), Lomborg does not identify a page number or either of the charts as the source of his claim that the Himalayan glaciers will run dry "toward the end of the century." Nor do the IPCC authors say anywhere in this section that the Himalayan glaciers will run dry toward the end of the century. However, the 2007 IPCC assessment report stated elsewhere: "Glaciers in the Himalaya are receding faster than in any other part of the world and, if the present rate continues, the likelihood of them disappearing by the year 2035 and perhaps sooner is very high if the Earth keeps warming at the current rate."[38] It is thus the case that the IPCC reported that, if the current rate of global warming continues, "the likelihood . . . is very high" that the Himalayan glaciers will disappear "by the year 2035 and perhaps sooner," and not "toward the end of the century," as Lomborg wrote.

Furthermore, in his reference Lomborg cited two sources to support his claim that the Himalayan glaciers will run dry "toward the end of the century." The first source was Section 3.4.1 in the 2007 IPCC assessment report (Working Group II), as noted above. Lomborg's second source was a 2003 study by Schneeberger et al. in the *Journal of Hydrology*.[39] However, the Schneeberger study, while its focus was the "mass balance of glaciers in the northern hemisphere," never mentions the Himalayas. And though the Schneeberger study lists seventeen glaciers and six regions that were studied, none are in the Himalayas.[40]

After misreporting the projected life expectancy of the Himalayan glaciers, as reported by the IPCC in 2007, and still writing clearly in the context of the Himalayas, Lomborg presented his view of the alleged benefits of the melting Himalayan glaciers: "Thus, global warming of glaciers means that a large part of the world can use more water for more than fifty years before they have to invest in extra water storage. These fifty-plus years can give the societies breathing space to tackle many of their more immediate concerns and grow their economies so that they will be better able to afford to build water-storage facilities."[41] Putting aside for now that water storage won't matter much if there is little water in the Himalayas to store after the glaciers have melted, the fifty-year framework of Lomborg's analysis is inapplicable to the Himalayan region, since the glaciers may disappear by 2035 (which is twenty-eight years, not fifty years, from the publication date of *Cool It*).

Nor is the solution to melting Himalayan glaciers as simple as "building water storage facilities," nor is it likely that most of those affected by melting Himalayan glaciers will "grow richer" as Lomborg suggests between now and when the glaciers melt. The 2007 IPCC assessment in fact undermines both of these contentions. Whereas Lomborg speaks of glacier-dependent rivers "actually *increas[ing]* their water contents" from the rapidly melting glaciers, and thus "providing *more* water to many of the poorest people in the world," leading to "a boon now,"[42] the IPCC described the impact of the rapidly melting glaciers in the Himalayas in decidedly less favorable terms:

- The entire Hindu Kush-Himalaya ice mass has decreased in the last two decades. Hence, water supply in areas fed by glacial melt water from the Hindu Kush and Himalayas, on which hundreds of millions of people in China and India depend, will be negatively affected.[43]
- The TAR [the 2001 IPCC assessment report] identified mountain regions as having experienced above-average warming in the 20th century, a trend likely to continue. Related impacts include an earlier and shortened snow-melt period, with rapid water release and downstream floods which, in combination with reduced glacier extent, could cause water shortage during the growing season. The TAR suggested that these impacts may be exacerbated by ecosystem degradation pressures such as land-use changes, over-grazing, trampling, pollution, vegetation destabilization and soil losses, in particular in highly diverse regions such as the Caucasus and Himalayas.[44]
- If current warming rates are maintained, Himalayan glaciers could decay at very rapid rates. Accelerated glacier melt would result in increased flows in some river systems for the next two to three decades, resulting in increased flooding, rock avalanches from destabilised slopes, and disruption of water resources. This would be followed by a decrease in flows as the glaciers recede. Permafrost degradation can result in ground subsidence [sinking], alter drainage characteristics and infrastructure stability, and can result in increased emissions of methane.[45]

Though the IPCC reported that "the impacts of climate change on water resources in Asia will be positive in some areas and negative in others,"[46] it was difficult to locate a description of positive changes in the Himalayan region, or elsewhere in Asia.[47] Thus, when Lomborg wrote that the region of the Himalayas will enjoy a fifty-year "breathing space" with "more water" and "a boon now" as a result of melting glaciers, he cited no references that in fact supported those claims.

Lomborg concluded his section on "Melting Glaciers" in *Cool It* with the statement, "While we often hear worries about how melting glaciers will lead to less water later, we seldom hear that it is a boon now." Not only is this likely untrue, given the IPCC analysis above, but neither of Lomborg's "boon now" and "water storage" analyses addresses the issue of a permanently lost, irreplaceable source of fresh water for hundreds of millions of people in Asia as a result of global warming. And though Lomborg himself provides no viable answers to the problem, he closes by ridiculing reductions in greenhouse emissions as a response to the problem of melting glaciers: "And when it is advocated that we instantly turn the big, hard knob of CO_2 cuts, which will do little to save the glaciers at very high costs, we should be asking whether there are other, nimbler, more efficient, and less expensive knobs to turn first, where we can help the world much more effectively."[48]

Rising Sea Levels

Lomborg's next alleged exposé—environmentalists' exaggerations about the threat of rising sea levels due to global warming—begins with the statement, "Another of the most doom-laden impacts from global warming is the rising sea levels." Lomborg then wrote: "This worry [of rising sea levels] is perhaps not surprising, since from time immemorial most cultures have had legends of catastrophic floods, which covered the entire Earth and left few animals and plants alive. In Western societies, the most famous version is the story of Noah saving what he could in his ark."[49]

While linking environmentalists' concerns about global warming and sea levels to our unconscious connections to ancient flood legends, Lomborg referenced this connection to a summary of flood stories from the *Encyclopedia Britannica*, including Noah's Ark; the Mesopotamian flood myth described in the *Epic of Gilgamesh;* the flood described by the Greek poet Pindar in the fifth century BC (when Zeus decided to destroy the Earth, saving only King Deucalion and his family); a flood myth described in India in 6th century BC about "Manu" (man) "who is warned by a fish about a coming flood"; and a Chinese flood myth about "a savior hero named Yü the Great."[50] It is these ancient legends, then, for Lomborg, more so than the sober evidence of the potential impact of global warming on sea levels, that have compelled environmentalists and scientists to express concerns and issue warnings.

Lomborg continued, "Many commentators powerfully exploit this biblical fear of flooding, as when Bill McKibben said of our responsibility for global warming that 'we are engaging in a reckless drive-by drowning of much of the rest of the planet and much of the rest of creation.'"[51] While the flood myths are fine, but misapplied here, Lomborg overlooked the fact that McKibben was addressing a "simple fact of physics," as McKibben put it, at the beginning of the article that Lomborg cited: "Warm water takes up more space than cold water does. That simple fact of physics, utterly inexorable, is one of the two or three most important pieces of information humans will have to grapple with in this century. And the people who get to grapple with it first are in places like Tuvalu, where suddenly the spring high tides are washing across the island of Funafuti, eroding foundations and salt-poisoning crops in the fields. Tuvalu is the canary in the miner's cage, and instead of choking it's drowning."[52]

McKibben wrote that "the warmer water, of course, is a product of steadily increasing global temperature, just like melting permafrost, shrinking glaciers, and increased evaporation over deserts." And global temperatures are rising "because we burn oil and coal and gas, which inexorably produce as a byproduct of their combustion carbon dioxide."[53] Though McKibben's analysis seems completely straightforward, to Lomborg he is somehow "exploiting" our "biblical fear of flooding."

Lomborg continues by noting that melting sea ice (as in Arctic sea ice)—that is, ice that sits on water rather than on land—does not contribute to rising sea levels. "Thus, contrary to common statements, the Arctic melting will not change sea levels," he concludes.[54] In this manner, Lomborg explodes a myth that doesn't exist in any significant sense. Nor does he identify who is allegedly making this supposedly "common" erroneous claim.

The next paragraph—the third in this section—is similarly weighted with misleading assertions. It begins: "In its 2007 report, the [IPCC] estimates that sea levels will rise about a foot over the rest of the century."[55] While this statement is not wholly inaccurate, it falls far short of summarizing the range of sea-level projections in the 2007

IPCC assessment report. For one thing, there was no single IPCC estimate projecting a one-foot sea-level rise over the course of this century. In a range of sea-level projections pursuant to six SRES development scenarios, the IPCC estimated that sea levels will rise anywhere from seven to twenty-three inches.[56] The higher sea-level projections are linked to development scenarios that will generate higher levels of greenhouse emissions for the twenty-first century. Thus, with only modest cuts in greenhouse emissions (which is Lomborg's policy prescription), sea levels could rise by over a foot to nearly two feet by the end of the century.

There are other key aspects of the IPCC's projections that Lomborg did not mention in any detail. For example, there is much uncertainty with respect to projections of sea-level rise, as the 2007 IPCC assessment conceded about its own projections: "The [IPCC] sea level projections do not include uncertainties in climate-carbon cycle feedbacks nor do they include the full effects of changes in ice sheet flow, because a basis in published literature is lacking. Therefore, the upper values of the ranges given are not to be considered upper bounds for sea level rise."[57]

In the same assessment report the IPCC also stated: "The [sea-level rise] projections include a contribution due to increased ice flow from Greenland and Antarctica at the rates observed for 1993–2003, but these flow rates could increase or decrease in the future. . . . Future changes in the Greenland and Antarctic ice sheet mass, particularly due to changes in ice flow, are a major source of uncertainty that could increase sea level rise projections."[58] The stipulations here that sea-level projections in the 2007 IPCC report were based on glacier flow rates on Greenland and Antarctica for 1993–2003, and that more recently observed changes in flow rates could have the effect of increasing sea-level projections, are crucial qualifications concerning the IPCC's projections for sea-level rise this century. This is because there is significant evidence—reported from 2002 to the present—that glacier flow rates have indeed dramatically changed in recent years.

Recall that a glacier is "a mass of land ice." When glaciers melt, or break up and slide into the ocean, they contribute to sea-level rise because there is a transfer of ice (water) from the land to the ocean. This happens in part, according to the IPCC, because glaciers "flow downhill under gravity" toward the ocean by "sliding at the base," where the bottom of the ice mass meets the land (bedrock).[59] An important study published in *Science* in 2002 found that, in the context of warming temperatures, "glacial sliding is enhanced by rapid migration of surface meltwater to the ice-bedrock interface";[60] that is, it lubricates the ice-land interface, and accelerates the flow of ice into the ocean. This study (Zwally et al., 2002) was among the first published reports to alert the scientific community to major changes in the flow rates of glaciers on the Greenland ice sheet, while also describing the mechanism by which this was occurring. The concern is that the massive Greenland ice sheet could break up at a faster rate than previously assumed, and thus increase sea levels beyond earlier projections. The same *Science* study reported these implications: "The interaction among warmer summer temperatures, increased surface meltwater production, water flow to the base, and increased basal sliding provides a mechanism for rapid response of the ice sheets to climate change. In general, a direct coupling between increased surface melting and ice-sheet flow has been given little or no consideration in estimates of ice-sheet response to climate change."[61]

Shortly after this study was published, several studies reported increases in glacier flows in Greenland. Unfortunately, and as it conceded, the 2007 IPCC assessment did not fully consider the implications of these studies when it calculated its sea-level projections for the twenty-first century. For example, citing measurements conducted by scientists from the University of Maine, the British newspaper *Independent* reported in July 2005:

> Scientists monitoring a glacier in Greenland have found it is moving into the sea three times faster than a decade ago. Satellite measurements of the Kangerdlugssuaq glacier show that, as well as moving more rapidly, the glacier's boundary is shrinking dramatically—probably because of melting brought about by climate change. The Kangerdlugssuaq glacier on Greenland's east coast is one of several that

> drains the huge Greenland ice sheet. The glacier's movements are considered critical in understanding the rate at which the ice sheet is melting.
>
> Kangerdlugssuaq is about 1,000 meters (3,280ft) thick, about 4.5 miles wide, extends for more than 20 miles into the ice sheet and drains about 4 per cent of the ice from the Greenland ice sheet. Experts believe any change in the rate at which the glacier transports ice from the ice sheet into the ocean has important implications for increases in sea levels around the world. If the entire Greenland ice sheet were to melt into the ocean it would raise sea levels by up to seven meters (23ft), inundating vast areas of low-lying land, including London and much of eastern England.
>
> Computer models suggest that this would take at least 1,000 years but even a sea-level rise of a meter would have a catastrophic impact on coastal plains where more than two-thirds of the world's population lives.[62]

The *Independent* also reported that measurements of Greenland's Kangerdlugssuaq glacier showed that it had moved at a rate of 3.1 to 3.7 miles per year in 1988 to 1996, and by 2005 was moving at a rate of 8.7 miles a year. Gordon Hamilton, the scientist who directed the study, reported that "Kangerdlugssuaq is probably the fastest-moving glacier in the world." Hamilton also stated: "This is a dramatic discovery. There is concern that the acceleration of this and similar glaciers and the associated discharge of ice [into the ocean] is not described in current ice-sheet models of the effects of climate change."[63]

Seven months later, in February 2006, the *Independent* reported again on Greenland's glaciers from a conference of the American Association for the Advancement of Science. The *Independent's* report is quoted here at length:

> Global warming is causing the Greenland ice cap to disintegrate far faster than anyone predicted. A study of the region's massive ice sheet warns that sea levels may—as a consequence—rise more dramatically than expected.
>
> Scientists have found that many of the huge glaciers of Greenland are moving at an accelerating rate—dumping twice as much ice into

the sea than five years ago—indicating that the ice sheet is undergoing a potentially catastrophic breakup.

The implications of the research are dramatic given Greenland holds enough ice to raise global sea levels by up to 21ft, a disaster scenario that would result in the flooding of some of the world's major population centres, including all of Britain's city ports.

Satellite measurements of the entire land mass of Greenland show that the speed at which the glaciers are moving to the sea has increased significantly over the past 10 years with some glaciers moving three times faster than in the mid-1990s.

Scientists believe that computer models of how the Greenland ice sheet will react to global warming have seriously underestimated the threat posed by sea levels that could rise far more quickly than envisaged.

The latest study, presented at the American Association for the Advancement of Science, in St Louis, shows that rather than just melting relatively slowly, the ice sheet is showing all the signs of a mechanical break-up as glaciers slip ever faster into the ocean, aided by the "lubricant" of meltwater forming at their base.

Eric Rignot, a scientist at the Jet Propulsion Laboratory and the California Institute of Technology in Pasadena, said that computer models used by the UN's International Panel on Climate Change have not adequately taken into account the amount of ice falling into the sea from glacial movements.

Yet the satellite study shows that about two-thirds of the sea-level rise caused by the Greenland ice sheet is due to icebergs breaking off from fast-moving glaciers rather than simply the result of water running off from melting ice.

"In simple terms, the ice sheet is breaking up rather than melting. It's not a surprise in itself but it is a surprise to see the magnitude of the changes. These big glaciers seem to be accelerating, they seem to be going faster and faster to the sea," Dr Rignot said.

"This is not predicted by the current computer models. The fact is the glaciers of Greenland are evolving faster than we thought and the models have to be adjusted to catch up with these observations," he said.[64]

On the same day it published this report (February 17, 2006), the *Independent* also published a guest column by NASA scientist James

Hansen, who assessed the significance of these scientific findings concerning Greenland's glaciers:

> A satellite study of the Greenland ice cap shows that it is melting far faster than scientists had feared—twice as much ice is going into the sea as it was five years ago. The implications for rising sea levels—and climate change—could be dramatic. . . .
>
> This new satellite data is a remarkable advance. We are seeing for the first time the detailed behavior of the ice streams that are draining the Greenland ice sheet. They show that Greenland seems to be losing at least 200 cubic kilometers of ice a year. It is different from even two years ago, when people still said the ice sheet was in balance. . . .
>
> Our understanding of what is going on is very new. Today's forecasts of sea-level rise use climate models of the ice sheets that say they can only disintegrate over a thousand years or more. But we can now see that the models are almost worthless. They treat the ice sheets like a single block of ice that will slowly melt. But what is happening is much more dynamic.[65]

Hansen, one of the premier climate scientists in the United States, explained the potential implications of the faster glacier flows in Greenland:

> How fast can this go? Right now, I think our best measure is what happened in the past. We know that, for instance, 14,000 years ago sea levels rose by 20m in 400 years—that is five meters in a century. This was towards the end of the last ice age, so there was more ice around. But, on the other hand, temperatures were not warming as fast as today.
>
> How far can it go? The last time the world was three degrees warmer than today—which is what we expect later this century—sea levels were 25m higher. So that is what we can look forward to if we don't act soon [to reduce greenhouse emissions]. None of the current climate and ice models predict this. But I prefer the evidence from the Earth's history and my own eyes. I think sea-level rise is going to be the big issue soon, more even than warming itself.[66]

Hansen observed that "it's hard to say what the world will be like if this happens. It would be another planet."

In August 2006, and citing a study published in *Science*, the *San*

Francisco Chronicle reported: "The vast ice cap that covers Greenland nearly three miles thick is melting faster than ever before on record, and the pace is speeding year by year," while "the consequence is already evident in a small but ominous rise in sea levels around the world, a pace that is also accelerating." The *Chronicle* reported that the University of Texas research team (authors of the 2006 study published in *Science*) found that "surface melting of Greenland's ice cap reached 57 cubic miles a year between April of 2002 and November 2005, compared to about 19 cubic miles a year between 1997 and 2003." It quoted one of the scientists: "The sobering thing is to see that the whole process of glacial melting is stepping up much more rapidly than before."[67]

The *Chronicle*, referring also to a study on West Antarctica's ice sheet by physicists from the University of Colorado, reported how these findings on Greenland and West Antarctica would affect the IPPC's sea-level projections:

> A recent report from the Intergovernmental Panel on Climate Change —known as the IPCC—estimated that during all of the past century worldwide melting ice from global warming had raised sea levels by only two-tenths of a millimeter a year, or about 20 inches for the entire century.
>
> But, according to [Jianli] Chen and his Texas team, the melting of Greenland's ice cap is already raising global sea levels by six-tenths of a millimeter a year, and the Colorado group estimates that melting of the West Antarctica Ice Sheet alone is adding up to four-tenths of a millimeter of fresh water to sea levels each year. In other words, the global sea level, due to melting of the ice in Greenland and Antarctica combined, is already rising 10 times faster than the IPCC's tentative estimates, the two analyses indicate.[68]

All of these studies and news articles were available to Lomborg while he was writing *Cool It*, but he took little or none of it into account: the 2007 IPCC assessment that had used glacier-flow rates from 1993 to 2003 in its sea-level projections; studies published in 2002 and afterward that reported greatly accelerated glacier-flow rates; and the acknowledgment in the 2007 IPCC assessment itself that it did not fully consider the impact of the newly reported glacier flow rates on sea-level projections. Thus, there is questionable value to Lomborg's narrowly accurate assertion that, "in its 2007 report, the UN estimates that sea levels will rise about a foot over the rest of the century." Nevertheless, Lomborg's highly uncertain estimate anchors his argument that environmentalists exaggerate the threat of sea-level rise from global warming.

Also, after asserting that the 2007 IPCC report projected a one-foot increase in sea level by the end of the twenty-first century, Lomborg wrote: "While this is not a trivial amount, it is also important to realize that it is certainly not outside historical experience. Since 1860, we have experienced a sea-level rise of about a foot, yet this has clearly not caused major disruptions."[69] This statement is likewise problematic, since Lomborg's claim that sea levels in the twenty-first century will rise within a relatively benign historical range depends entirely on his decision to move the historical goal posts back only as far as the mid-1800s. If Lomborg, on the other hand, had decided to look back about 14,000 years, he could have reported, as Hansen did, that sea levels rose by twenty meters in four hundred years (five meters per century) when temperatures were as warm as they are projected to be at some point this century. In addition, Lomborg neglected to mention that sea levels rose more in the twentieth century than in the previous two thousand years, and will rise at a faster rate in the twenty-first century, as the 2007 IPCC assessment report noted: "There is strong evidence that global sea level gradually rose in the 20th century and is currently rising at an increased rate, after a period of little change between AD 0 and AD 1900. Sea level is projected to rise at an even greater rate in this century."[70]

Lomborg next criticizes a 2001 report about global warming by *U.S. News and World Report*,[71] which, according to Lomborg, exaggerated the projected impact of sea-level rise. *U.S. News* wrote, as quoted by Lomborg, that "by mid-century, the chic Art Deco hotels that now line Miami's South Beach could stand waterlogged and abandoned." Lomborg cited this sentence as an example of how "the risk of sea-level rise is strongly dramatized in the public discourse." He then wrote: "Yet sea-level increase by 2050 will be about five inches—no more than the change we have experienced since 1940 and less than the change those Art Deco hotels have already stood through."[72] The fact is that Lomborg doesn't know with any such exactitude ("about five inches") how much sea levels will rise by 2050. In fact, his only source to support this claim is an article in a defunct travel magazine (*Travel Holiday*) that makes no mention of sea-level rise or global warming.[73]

Furthermore, in criticizing the comments by *U.S. News* about the Art Deco hotels on Miami Beach, Lomborg ignored the 2001 IPCC assessment report conclusions about the threat to U.S. coastlines, including Florida, from global warming and rising sea levels. This is an especially relevant omission, since the late January 2001 report by *U.S. News* in fact was about the newly released 2001 IPCC assessment report. In its 2001 assessment, the IPCC reported with respect to sea-level rise that "the greatest vulnerability is expected in areas that recently have become much more developed, such as Florida and much of the U.S. Gulf and Atlantic coasts."[74] In light of these remarks, and because it is difficult to imagine an area that is as highly developed and as close to the ocean as Miami Beach, Florida, including the Art Deco hotels, it seems that *U.S. News* chose an entirely appropriate U.S. symbol to illustrate what the IPCC had just reported at the time.

Lomborg's attacks on generally accurate claims about global warming, as in the case of *U.S. News* above, are common in *Cool It*. And Al Gore's *An Inconvenient Truth*—both the documentary film and book—is a frequent target. In one such attack on Gore about the issue of global warming and sea-level rise, Lomborg again referenced his unsupported claim of a five-inch sea-level increase by year 2050, and wrote:

> [F]ive inches will simply not leave Miami Beach hotels waterlogged and abandoned. But this of course is exactly the opposite of what we often hear. In a very moving section of the film *An Inconvenient Truth*, we see how large parts of Florida, including all of Miami, will be inun-

> dated by twenty feet of water. We see equally strong clips of San Francisco Bay being flooded, the Netherlands being wiped off the map, Beijing and then Shanghai being submerged, Bangladesh being made uninhabitable for sixty million people, and the deluging of even New York City and its new World Trade Center memorial.
>
> How is it possible that one of today's strongest voices on climate change [Gore] can say something so dramatically removed from the best science? The IPCC estimates a foot, Gore tops them twenty times. Well, technically, Al Gore is not contradicting the UN, because he simply says: "If Greenland melted or broke up and slipped into the sea—or if half of Greenland and half of Antarctica melted or broke up and slipped into the sea, sea levels worldwide would increase by 18 and 20 feet." He is simply positing a hypothetical and then in full graphic and gory detail showing us what—hypothetically—would happen to Miami, San Francisco, Amsterdam, Beijing, Shanghai, Dhaka, and then New York City.[75]

An Inconvenient Truth (the film and the book) was issued in 2006 and thus *before* the 2007 IPCC assessment report, and *Cool It* was published in late 2007 and thus *after* the 2007 IPCC assessment was issued in early 2007; consequently Lomborg, not Gore, had access to the 2007 IPCC projections of sea-level rise. Given this context, note that Lomborg is comparing the alleged one-foot sea-level rise projection from the 2007 IPCC assessment report to Gore's 2006 book and movie. Thus, when Lomborg calls Gore's assertion into question—"How is it possible that one of today's strongest voices on climate change can say something so dramatically removed from the best science?"—Lomborg is holding Gore accountable to the 2007 IPCC estimate that had not yet been issued when *An Inconvenient Truth* was released in 2006. The only other "best science" that Lomborg cited as being at odds with Gore's account was Lomborg's claim that sea levels would rise by five inches by 2050, though Lomborg referenced this claim to a travel magazine.

As stated previously, the 2007 IPCC assessment did not simply project a one-foot sea-level increase by year 2100; it presented a range of possible increases up to nearly two feet. And the 2007 IPCC assessment stated that its sea-level projections were uncertain and could change pursuant to changing conditions, including increasing glacier flow rates, which is now a major scientific concern. In fact, the 2007 IPCC chart that projected sea-levels for the twenty-first century (and that Lomborg cited while issuing his estimate of a one-foot sea-level increase this century) stipulated that the projections "exclude future rapid dynamical changes in ice flow" — that is, changes that almost certainly have already occurred, as reported in several studies issued since 2002. So while Lomborg confidently reduces the IPCC's tentative 2007 sea-level projections for the twenty-first century to one foot, he omits these important qualifying facts. Thus, Lomborg's criticism of Gore is dubiously framed before he proceeds to pursue his discussion of the merits of Gore's arguments about global warming and sea levels.

Furthermore, in *An Inconvenient Truth* — in both the film and book — the sea-level changes depicted for Florida, Manhattan, the San Francisco Bay area, the Netherlands, China, India, and Bangladesh were all clearly shown in the context of a future partial or total loss of the Greenland and West Antarctica ice sheets, as Lomborg indicated, but were less hypothetical than Lomborg alleged. In Gore's book, the sea-level depictions of these places were directly preceded by eight pages of discussion pertaining to the ultimate fate of the ice sheets of Greenland and Antarctica if greenhouse-gas-driven global warming isn't checked.[76] Gore began this section, "Now consider the much larger areas of ice in Antarctica and Greenland that are at risk."[77] Gore then presented a brief summary of recent discoveries pertaining to Greenland and Antarctica, including glacier surface melting and glacier flow rates.[78] Gore did not, as Lomborg attempted to imply, situate this discussion of what might happen to Florida, Bangladesh, and the other locations over the next one hundred years; rather, his analysis, though presented with a sense of urgency, followed no chronology, and instead sought to illustrate what could occur if the Greenland and West Antarctic ice sheets broke up and melted — a not entirely hypothetical possibility, it indeed must be taken into account in any serious discussion of global warming and sea-level rise.[79] And Gore's rendition of sea-level rise in the context of the uncertain but increasingly worrisome fate of the Greenland and West Antarctica ice sheets was directly preceded by the comments of British scientist David King, who, while addressing "the potential consequences of large changes" in the Greenland and West Antarctica ice sheets, stated: "The maps of the world will have to be redrawn."[80]

Finally, Gore's analysis was also consistent with the commentary by James Hansen cited earlier. Writing in 2006 about "the implications for rising sea levels" of newly published research concerning the ice dynamics on Greenland, Hansen observed that "you could imagine great armadas of icebergs breaking off Greenland and melting as they floated south," and referred to "huge areas being flooded."[81] Hansen then concluded, like Gore, with a sense of urgency: "How long have we got? We have to stabilize emissions of carbon dioxide within a decade, or temperatures will warm by more than one degree. That will be warmer than it has been for half a million years, and many things could become unstoppable. If we are to stop that, we cannot wait for new technologies like capturing emissions from burning coal. We have to act with what we have. This decade, that means focusing on energy efficiency and renewable sources of energy that do not burn carbon. We don't have much time left."[82]

Perhaps Lomborg genuinely believes that Gore's and Hansen's arguments are "hypothetical." But as this chapter has clearly shown, Lomborg joins neither Gore nor Hansen in the global warming debate on a level in which both sides reason by the same rules of evidence.

six
ON GREENLAND AND THE MISSING FIGURES

Because of the importance of sea-level rise as a likely major impact of global warming (as described in the previous chapter), and given that Lomborg's estimate of a mere one-foot sea-level increase by the end of the century is a major component of Lomborg's Theorem, it is worth investigating further how Lomborg arrived at that estimate. Lomborg began with another of many references to Al Gore and *An Inconvenient Truth:* "Gore is correct in identifying Antarctica and Greenland as the most important players if he is to support his hypothetical twenty feet. The UN estimates that over the century by far the largest contribution to sea-level rise will be warmer water expanding—this alone will constitute nine of the almost twelve inches by 2100."[1]

In response, it is worth pointing out that the second sentence of this passage is referenced to the 2007 IPCC assessment report. To be precise, it is referenced to "fig. 10.6.1" (Figure 10.6.1) in Working Group I (WGI) of the 2007 IPCC assessment report. But no such figure can be found, and Lomborg's reference provides no page numbers in which to narrow the search. Although the contribution of thermal expansion of the oceans from global warming to sea-level rise is an uncontroversial issue, Lomborg's precise estimate of a nine-inch sea-level increase by year 2100 due to thermal expansion is an interesting assertion. Unfortunately, Lomborg's source—Figure 10.6.1—cannot be located in the 2007 IPCC assessment report.

Let's assume in this instance that Lomborg inadvertently mislabeled Section 10.6.1 of the report as Figure 10.6.1. In fact, Section 10.6.1 is titled, "Global Average Sea Level Rise Due to Thermal Expansion."[2] Section 10.6.1 contains "Figure 10.31," which depicts three projections (using three SRES scenarios) of thermal expansion of the world's oceans to year 2100 (and thus not a single projection). The first SRES scenario (A1B) projected a thermal expansion of the world's oceans of 0.14 to 0.38 meters by year 2100; the second SRES scenario (A2) projected a thermal expansion of 0.15 to 0.36 meters by 2100; and the third SRES scenario (B1) projected a thermal expansion of 0.1 to 0.29 meters. These three projections thus represent a range of thermal expansion by year 2100 of 0.1 to 0.38 meters, or 4 to 15 inches. Though Lomborg's projection of a nine-inch thermal expansion of sea level by year 2100 represents a rough median within this range, the ranges themselves are at odds with his singularly precise estimate.

In Lomborg's next sentence — that is, after claiming that thermal expansion "will constitute nine of the almost twelve inches [of sea-level increase] by 2100," (for which "fig. 10.6.1" is supposed to be the source) — he wrote: "Melting glaciers and ice caps will contribute a bit more than three inches over the century."[3] Lomborg referenced this assertion to "fig. 10.6.3" in WGI of the 2007 IPCC assessment. As in the previous instance, there is no Figure 10.6.3. However, there is a "Section 10.6.3," titled "Glaciers and Ice Caps."[4] But Section 10.6.3 consists of only three sentences, none of which support the claim that "melting glaciers and ice caps will contribute a bit more than three inches." Digging further, we can locate Section 10.6.3.1 ("Mass Balance Sensitivity to Temperature and Precipitation"), Section 10.6.3.2 ("Dynamic Response and Feedback on Mass Balance"), and Section 10.6.3.3 ("Glaciers and Ice Caps on Greenland and Antarctica").[5] But these sections are filled with dense calculations and citations across three pages, with no apparent substantiation of Lomborg's claim that melting glaciers and ice caps will contribute a bit more than three inches to sea-level rise by year 2100. And since Lomborg did not actually reference these sections, there is no specified page or section number to look for.

Also, immediately after referencing "fig. 10.6.3" in his endnote (page 180 in *Cool It*) as the source for his claim that "melting glaciers and ice caps will contribute a bit more than three inches over the century" to sea-level rise, Lomborg wrote (and referring to his estimate of "a bit more than three inches"): "It is actually 8.8 centimeters [3.5 inches], but the 0.8 seems to get lost somewhere in the sums."

Not only is there no reference to 8.8 centimeters in Figure 10.6.3, because there is no such figure, but there is no mention of 8 or 8.8 centimeters in the sections listed above, or anywhere in the chapter of the IPCC assessment report that Lomborg references.

After writing that melting glaciers and ice caps will lead to a three-inch increase in sea-levels by 2100, Lomborg wrote: "Likewise, Greenland is expected to contribute 1.4 inches by itself. This adds up to 13.5 inches over the coming century."[6] Lomborg referenced this claim to "fig. 10.6.4" in Working Group I of the 2007 IPCC assessment report; however, like Figure 10.6.1 and Figure 10.6.3, there is no Figure 10.6.4. There is a "Section 10.6.4" titled "Ice Sheets,"[7] with subsections on the "Surface Mass Balance" and "Dynamics" of the Greenland and Antarctic ice sheets that span five pages.[8] But they do not report that Greenland is expected to contribute 1.4 inches to sea-level rise by the end of the century.

Lomborg's pattern of referencing nonexistent IPCC figures to support a core plank of Lomborg's Theorem—that global warming will cause no catastrophic sea-level rise—continued in the next sentence: "However, as the world warms, Antarctica will not noticeably start melting (it is still way too cold), but because global warming also generally produces more precipitation Antarctica will actually be accumulating ice, *reducing* sea levels by two inches."[9] To support this assertion, Lomborg again cited the nonexistent Figure 10.6.4 (Section 10.6.4 did not report that Antarctica will contribute a net two inches of sea-level reduction).

After providing nonexistent referenced sources to substantiate a foundational claim in *Cool It*—that global warming will cause sea levels to rise by only one foot by the end of this century—Lomborg concluded: "Thus, the total estimate of about one foot."[10] Thus, Lomborg referenced only these IPCC figures to itemize his assertions of a one-foot sea-level rise, even though none of these sources can be found in the 2007 IPCC assessment report.

Writing about Greenland, Lomborg continued his questionable assertions. For example, one page after describing Gore's projections of possible sea-level rise as merely "hypothetical," Lomborg wrote: "Greenland is, as the IPCC pointed out, experiencing a small overall mass loss. Some analyses have shown more rapid loss in recent years (2002–5), but by early 2007 two of the major glaciers in Greenland were again seen reverting to much lower rates of ice-mass loss. Even with the most extreme estimates of Greenland melting over a couple of years, a sea-level rise of twenty feet would take one thousand years. In a recent overview of all the major models of sea-level increase, Greenland's contribution over the coming century is at most two inches. Some even posit a tiny decrease in sea levels from increased snow outweighing the melting of Greenland's glaciers."[11]

Lomborg cited eleven studies in support of this argument. However, with one exception (Oerlemans et al., 2005),[12] which is referenced to the assertion that "Greenland's contribution over the coming century is at most two inches," Lomborg provided no page numbers to the other ten studies that would allow the reader to locate the alleged factual basis for the other assertions; Lomborg simply leaves it to the reader to sort blindly through this multitude of references. And, a few of those assertions are of questionable merit. For example, Lomborg wrote: "But by early 2007 two of the major glaciers in Greenland were again seen reverting to much lower rates of ice-mass loss." This claim appears designed to offset the preceding findings that Greenland was experiencing "a small overall mass loss," and a "more rapid loss in recent years." However, as James Hansen pointed out: "That a glacier on Greenland slowed after speeding up, used as 'proof' that reticence [about ice-sheet loss] is appropriate, is little different than the common misconception that a cold weather snap disproves global warming."[13]

Furthermore, a quick chronological review of Lomborg's own referenced sources do not appear to support the above claim that Greenland's impact on sea levels by the end of the century will range from a net negative contribution to a net increase of two inches. One such source, a paper published in *Nature* in September 2006 titled, "Acceleration of Greenland Ice Mass Loss in Spring 2004," reported a measured increased rate of loss in the Greenland ice sheet:

> In 2001 the Intergovernmental Panel on Climate Change projected the contribution to sea level rise from the Greenland ice sheet to be between -0.02 and +0.09 m [-0.8 to 3.5 inches] from 1990 to 2100. However, recent work has suggested that the ice sheet responds more quickly to climate perturbations than previously thought, particularly near the coast. . . . The rate of loss increased by 250 percent between the periods April 2002 to April 2004 and May 2004 to April 2006, almost entirely due to accelerated rates of ice loss in southern Greenland.[14]

This report not only indicates that the 2001 IPCC assessment projected that Greenland's contribution to sea levels would range from minus 1 inch to 3.5 inches, which thus shows a high-end contribution of 3.5 inches—more than Lomborg's "at most two inches"—it also concludes that "the rate of [ice mass] loss [on Greenland] increased by 250 percent between the periods April 2002 to April 2004 and May 2004 to April 2006." Thus, one might expect that Greenland's contribution to sea levels by the end of the century would be greater than what the IPCC had projected in 2001.

Other studies also referenced by Lomborg (within the same group of studies) similarly reported that Greenland's loss of ice mass had dramatically increased since 2001. A study published in *Science*, also in September 2006, titled, "Satellite Gravity Measurements Confirm Accelerated Melting of Greenland Ice Sheet," reported that "the estimated total ice melting rate over Greenland" is 239 cubic kilometers per year, "mostly from East Greenland," and that this estimate "agrees remarkably well with a recent estimate" of 224 cubic kilometers per year. The study reported "accelerated melting since the summer of 2004, consistent with the latest remote sensing measurements."[15]

In November 2006, *Science* published another study on Greenland's ice-mass loss, also referenced by Lomborg, which reported that from 2003 to 2005, the ice sheet lost 101 gigatons a year (1 gigaton equals 1 billion tons), and that "the overall rate of loss reflects a considerable change in trend from a near-balance during the 1990s but is smaller than some recent estimates." The study concluded: "Our new results suggest that the processes of significant ice depletion at the margins, through melting and glacier acceleration, are beginning to dominate the interior growth [of the ice mass] as climate warming has continued."[16]

Another study referenced by Lomborg, published in *Science* in March 2007, reported: "After a century of polar exploration, the past decade of satellite measurements has painted an altogether new picture of how Earth's ice sheets are changing. As global temperatures have risen, so have rates of snowfall, ice melting, and glacier flow. Although the balance between these opposing processes has varied considerably on a regional scale, data show that Antarctica and Greenland are each losing mass overall."[17]

While Lomborg represented these studies as evidence of "a small overall mass loss" on Greenland, with some studies "showing more rapid loss in recent years," they in fact reported *higher* rates of ice-mass loss and net contributions to sea-level rise. In addition, while he cited Oerlemans et al., 2005, which projected a small contribution to sea-level rise from Greenland by the end of the century ("at the most two inches"), Lomborg did not mention the first two sentences of the Oerlemans study abstract: "In this paper, we report on an approach to estimate the contribution of Arctic glaciers to sea-level change. In our calculation we assume that a static approach is feasible."[18] This appears to indicate that Lomborg gave the most weight to an estimate of the contribution to sea levels from land ice in the Arctic (which predominately features the Greenland ice sheet) to a 2005 study whose methods did not reflect the dynamic changes to Greenland's ice sheet that had been widely reported—if not fully by 2005, then certainly by 2006 and 2007—in studies that Lomborg himself referenced.

Immediately following his claim that Greenland would contribute "at most two inches" to sea-level rise by the end of the century—while citing "a recent overview of all the major models of sea-level increase" (Oerlemans et al., 2005)—Lomborg pointed similarly to a later "overview" (Gregory and Huybrechts, 2006): "In another overview, all models clearly show both Greenland and Antarctica making small contributions [to sea-level rise] over the century."[19] But Lomborg did not mention important limitations acknowledged in the study itself:

> In presently available continental ice-sheet models, the dynamical response has a time-scale of centuries. Recent observations of accelera-

> tion of glaciers behind the Larsen-B ice shelf [on the Antarctic Peninsula], of ice streams in the Amundsen Sea sector of West Antarctica, and of many Greenland outlet glaciers suggest that the time-scales for ice-dynamical changes may be much shorter. Such accelerated flow leads to increased ice discharges into the ocean, but the relevant dynamical processes are not properly understood nor included in continental ice-sheet models. . . . This therefore represents an important uncertainty for predictions of sea level, but one which is beyond the scope of this paper to address.[20]

Thus both "overviews" relied upon by Lomborg (Oerlemans et al., 2005; Gregory and Huybrechts, 2006) were arguably dated by 2007 – including, as Gregory and Huybrechts reported above – by "accelerated [glacier] flows," "increased ice discharges," and "dynamical processes" that went "beyond the scope" of their paper.

Immediately following his references to these two studies (Oerlemans et al., 2005; Gregory and Huybrechts, 2006), and apparently determined to convince his readers that sea-level rise due to global warming will be no catastrophe, Lomborg wrote: "The IPCC estimates that the very worst additional increase to be expected from Greenland could be eight inches over the century, but this is possible only in a model where CO_2 levels rise two to four times more than expected by 2100. Thus, there is very little support for the assumption of a twenty-foot sea rise."[21] As before, Lomborg referenced this passage to a figure ("fig. 10.6.4.3") that does not exist in the 2007 IPCC assessment report.

Lomborg's presentation of sea-level rise, and climate change in general, *at best* describes only one side of the story. Both sides are illustrated by the following passage from the 2007 IPCC assessment report, wherein the word "However" (italics added) marks the division between the two sides. The first part of the IPCC assessment roughly comports with Lomborg's version, while the second half reflects scientific concerns that the 2007 IPCC projections pertaining to climate change (and sea-level rise) are too conservative:

> Abrupt climate changes, such as the collapse of the West Antarctica Ice Sheet, the rapid loss of the Greenland Ice Sheet or large-scale changes of ocean circulation systems, are not considered likely to occur in the 21st century, based on currently available model results.
>
> *However*, the occurrence of such changes becomes increasingly more likely as the perturbation of the climate system progresses.
>
> Physical, chemical and biological analyses from Greenland ice cores, marine sediments from the North Atlantic and elsewhere and many other archives of past climate have demonstrated that local temperatures, wind regimes and water cycles can change rapidly within just a few years. The comparison of results from records in different locations of the world shows that in the past major changes of hemispheric to global extent occurred. This has led to the notion of an unstable past climate that underwent phases of abrupt change. Therefore, an important concern is that the continued growth of greenhouse gas concentrations in the atmosphere may constitute a perturbation sufficiently strong to trigger abrupt changes in the climate system. Such interference with the climate system could be considered dangerous, because it would have major global consequences.[22]

Lomborg essentially related the first part of this IPCC-reported scenario, and ignored the second part.

Someone reading the IPCC summaries and the IPCC reservations about its own projections of sea-level rise might ask whether the 2007 IPCC assessment should be viewed as the authoritative source for projected sea levels in ensuing years. This is an important question, since the IPCC's sea-level projections apparently did not incorporate the sea-level implications of the studies briefly summarized above, which militate against even the IPCC's upper-range projection of a sea level rise of nearly two feet. As even the mere titles of these studies suggest, this would seem to be a serious question with respect to the IPCC's projections: "Surface Melt-Induced Acceleration of Greenland Ice-Sheet Flow" (*Science*, 2002),[23] "Mass Changes of the Greenland and Antarctic Ice Sheets and Shelves and Contributions to Sea-Level Rise: 1992–2002" (*Journal of Glaciology*, 2005),[24] "Acceleration of Greenland Ice Mass Loss in Spring 2004" (*Nature*, 2006),[25] "Satellite Gravity Measurements Confirm Accelerated Melting of Greenland Ice Sheet" (*Science*, 2006),[26] and "Recent Greenland Ice Mass Loss by Drainage System From Satellite Gravity Observations" (*Science*, 2006).[27]

Prior to these studies (which show evidence of rapid dynamic response of the ice sheets to climate change), the IPCC assessment report, issued in 2001, concluded that "accelerated sea level rise caused by rapid dynamic response of the ice sheets to climate change is very unlikely during the 21st century."[28] The studies cited above, published after 2001, weaken that conclusion. Likewise, the 2007 IPCC assessment report noted: "New evidence of recent rapid changes in the Antarctic Peninsula, West Antarctica and Greenland has again raised the possibility of larger dynamical changes in the future than are projected by state-of-the-art continental models [used in the 2001 and 2007 IPCC assessment reports] because these models do not incorporate all the processes responsible for the rapid marginal thinning currently taking place."[29] Thus, by its own reckoning, the IPCC's 2007 assessment did not fully consider the sea-level consequences of recently reported changes in the Greenland and Antarctic ice sheets noted in the studies listed above.

The IPCC's conservative sea-level projections were the subject of commentary by climate scientist James Hansen shortly after the 2007 IPCC assessment report was issued. The article, titled "Scientific Reticence and Sea Level Rise," referred to the prevailing caution about projecting sea-level rise, including in the 2007 IPCC assessment report, as Hansen wrote: "Caution, if not reticence, has its merits. However, in a case such as ice sheet instability and sea level rise, there is a danger in excessive caution. We may rue reticence, if it serves to lock in future disasters."[30]

Hansen questioned why the "concern about the danger of 'crying wolf' is more important than concern about the danger of 'fiddling while Rome burns.' "

Though he was writing at roughly the same time as Lomborg in 2007, with access to the same studies and data, Hansen chose a much different outlook. For example, Lomborg never competently addressed the issue of paleoclimate data and its relevance to the Greenland and Antarctic ice sheets and sea-level rise. In contrast, Hansen wrote that "extensive paleoclimate data confirm the common sense expectation that the net effect is for ice sheets to shrink as the world warms."[31] Paleoclimate is "climate during periods prior to the development of measuring instruments, including historic and geologic time, for which only proxy climate records [such as from ice cores] are available."[32] Referring again to paleoclimate data, Hansen wrote that the "global mean temperature three million years ago was only 2–3°C [3.6–5.4°F] warmer than today, while the sea level was 25 meters higher."[33]

Contingent on how much we reduce (or don't reduce) greenhouse emissions, the 2007 IPCC assessment report projects a range of temperature increases of 1.1–6.4°C (2–11.5°F) by the end of this century.[34] Thus, a middle-ground increase of 2–3°C (3.6–5.4°F) is well within the range of IPCC projections, if not assured, without an immediate and serious focus on reducing greenhouse emissions. In this context, Hansen wrote: "In assessing the likely effects of a warming of 3°C, it is useful to note the effects of the 0.7°C warming in the past century. This warming already produces large areas of summer melt on Greenland and significant melt on West Antarctica. Global warming of several more degrees, with its polar amplification, would have both Greenland and West Antarctica bathed in summer melt for extended melt seasons."[35]

Though Hansen is a prominent climate scientist, this commonsense analysis, albeit grounded in sophisticated data, is persuasive, and it exists in contrast with the highly questionable data in Lomborg's work. Furthermore, there appears to be no safe harbor in Lomborg's *Cool It* where this characterization of his data does not apply, including in a brief one-page sidebar concerning Antarctica's penguins, which I examine in the next chapter.

seven
THE PENGUINS SIDEBAR

While we transition in *Cool It* from Lomborg's analysis of the Greenland ice sheet to the West Antarctic ice sheet in the next chapter, it is worth visiting a small island, so to speak, in between—a sidebar in *Cool It* titled, "Penguins in Danger?" Lomborg begins by implying that Al Gore exaggerated the harmful impact of global warming, this time on Antarctica's penguins. Referring to Gore's *An Inconvenient Truth*, Lomborg wrote: "Al Gore also shows us how the rising temperatures on the Antarctic Peninsula have dramatically affected the emperor penguins, who were the subject of the 2005 documentary *March of the Penguins*. This colony of penguins, over five hundred yards from the pioneering French research station Dumont d'Urville, has been constantly monitored since 1952."[1]

Saying nothing further about Gore, the sidebar continued: "Its [the Emperor penguin] population was steady at around six thousand breeding pairs until the 1970s, when it dropped abruptly to about three thousand pairs, and it has remained stable since. This could possibly be linked to climate change, although the onetime decline makes it less likely."[2] The only source for this passage was a 2001 study published in *Nature* titled, "Emperor Penguins and Climate Change." Consistent with what Lomborg noted, the study in *Nature* reported that the Emperor penguin colony near Dumont d'Urville, studied from 1952 to 2000, "has declined by 50%." However, Lomborg did not mention that the *Nature* study attributed this decline to the penguins' "potential high susceptibility to climate change."[3] Instead, Lomborg wrote that the penguins' decline "could possibly be linked to climate change," although the "onetime decline," which lasted decades, "makes it less likely." This appears to be Lomborg's unmoored opinion, since the *Nature* study (which I will refer to again below) attributed the penguins' "potential high susceptibility" to global warming as a factor in their decline.

The fifth and sixth sentences of Lomborg's sidebar continue: "However, it [the Emperor penguin colony near the Dumont d'Urville station] is but a single, and rather small, colony out of about forty colonies around Antarctica. It is the best studied only because of its location."[4] Lomborg referenced these two sentences to an article that was on a Web page of the Australian Government Antarctic Division and titled, "Where Do They [Emperor Penguins] Breed?" In contrast to Lomborg's assertion that there are "about forty colonies [of Emperor penguins] around Antarctica," the Australian government's Web page provided a more nuanced estimate of the number of known Emperor penguin colonies: "Emperor penguin colonies occur around the Antarctic continent. In the past, over 30 colonies were sighted and a few were discovered only in the last 10 years"; "most colonies have not been visited for several decades"; and "many colonies are very difficult to get to."[5] Apparently based on this information Lomborg concluded that Emperor penguins have "about forty" colonies today, which would seem to be a high-end estimate. Lomborg did not include the caveats noted on the Australian government's Web site with his estimate.

The seventh sentence of Lomborg's sidebar reads: "Some of the largest colonies contain more than twenty thousand pairs each, several of which may be increasing."[6]

This statement, which is referenced to a 1997 paper in *Marine Ornithology*,[7] seems to be only partially accurate. Here is what the paper reported on the page that Lomborg footnoted: "The [Emperor penguin] colony at Point Géologie, monitored annually since the 1950s, which decreased substantially between the 1970s and late 1980s, has remained stable since then. The population size of the Auster colony has remained stable over the last eight years. Numbers of breeding pairs at Taylor Glacier are in close agreement with those obtained in the mid-1970s, indicating stability at this colony for at least 20 years. Colonies in the Ross Sea may be increasing currently."[8]

These assessments seem to indicate that Emperor penguin populations have been stable in three of the four areas observed, and that penguin colonies in *one* of the areas (and not "several," as Lomborg wrote) "may be increasing." Also, if Lomborg had broadened his focus in this instance beyond Emperor penguins, he could have reported additional information from the same page he was referencing:

> King Penguins *Aptenodytes patagonicus* are still increasing; except for Emperor *A. forsteri* and Gentoo *Pygoscelis papua* Penguins, all the other Antarctic and sub-Antarctic species (including Adélie *P. adeliae* and Chinstrap *P. antarctica* Penguins) are currently showing an overall decrease in populations compared with the situation a decade ago; this is also true for most regional populations. The situation is potentially most serious for the Macaroni Penguin *Eudyptes chrysolophus* and especially for the Rockhopper Penguin *E. chrysocome*, which is being recommended for Globally Threatened status in the next Red Data Book.[9]

Depending upon which species of penguins one selects to cite from the 1997 *Marine Ornithology* article, one could show either an increase in the penguin population (King penguins), no significant change (Emperor and Gentoo penguins), or a decline (Adélie, Chinstrap, Macaroni, and Rockhopper penguins). Lomborg chose to report only that the Emperor penguin population was increasing, even though the article he cited actually reported that the Emperor population showed no significant change.[10]

Lomborg's penguin sidebar contains nine sentences, the eighth of which reads: "The IUCN [International Union for the Conservation of Nature] estimates that there are almost two hundred thousand pairs [of Emperor penguins] and that the population is stable, placing it in the category 'least concern.' "[11] This is a true statement, as the IUCN reports on the page referenced by Lomborg: "The species is not believed to approach the thresholds for the population decline criterion of the IUCN Red List (i.e. declining more than 30% in ten years or three generations). For these reasons, the species is evaluated as Least Concern." The same IUCN page also reports that "global population trends [of Emperor penguins] have not been quantified, but populations appear to be stable." However, while Lomborg asserted that the IUCN estimated "almost two hundred thousand pairs" of Emperor penguins, the same IUCN Web site reports that the penguins have "a large global population estimated to be 270,000–350,000 individuals."[12] This would be, at most, 135,000 to 175,000 thousand pairs, not 200,000 pairs.

This particular IUCN Web page, which makes no mention of global warming, was not the only source that Lomborg could have cited to assess how the Emperor penguins are affected by global warming. For example, the 2007 IPCC assessment report noted a decline in Emperor penguin populations "that have experienced warming" (recall that "abundances" refer to population estimates): "Substantial evidence indicates major regional changes in Antarctic terrestrial and marine ecosystems in areas that have experienced warming. Increasing abundance of shallow-water sponges and their predators, declining abundances of krill, Adélie and Emperor penguins, and Weddell seals have all been recorded."[13]

Similarly, the 2001 study in *Nature* that Lomborg referenced, as noted above, described in some detail the susceptibility of Emperor penguins to global warming in regions of Antarctica that had experienced warming-related effects:

> We show that over the past 50 years, the population of emperor penguins (*Aptenodytes forsteri*) in Terre Adelie has declined by 50% because of a decrease in adult survival during the late 1970s. At this time there was a prolonged abnormally warm period with reduced sea-ice extent. Mortality rates increased when warm sea-surface temperatures occurred in the foraging area and when annual sea-ice extent was reduced, and were higher for males than for females. . . . These results indicate strong and contrasting effects of large-scale oceanographic processes and sea-ice extent on the demography of emperor penguins, and their potential high susceptibility to climate change.[14]

Though Lomborg cited this *Nature* study to report that "the onetime decline" in the Emperor penguin population, which lasted fifty years, "makes it less likely" that the decline was the result of global warming, the same study, which was the only source Lomborg cited for the passage containing this claim, reported that the 50 percent decline over fifty years coincided with "a prolonged abnormally warm period with reduced sea-ice extent," that "mortality rates increased when warm sea-surface temperatures occurred in the foraging area and when annual sea-ice extent was reduced," and that these results indicate the Emperor penguins' "potential high susceptibility to climate change."

The ninth and last sentence of Lomborg's sidebar states: "Moreover, the other main Antarctic penguin, the Adélie, has in the same region seen an increase of more than 40 percent over the past twenty years, underscoring the problem in simply blaming global warming and not telling the full story."[15] In this instance Lomborg accurately cited a 2006 study published in *Polar Biology*, which reported that the Adélie penguin colony in Pointe Géologie Archipelago increased "1.77% per year" between 1984 and 2003.[16] Also, this study complements a report by David Ainley, a leading researcher on Adélie penguins, who found that climate change, at least in the short term, has contributed to an increase in the penguins' numbers on Ross Island in Antarctica's Ross Sea.[17] At the same time, however, and to illustrate the variability in the data on penguins and global warming, a World Wildlife Fund study, issued in December 2007, shortly after *Cool It* was published, reported different results about the impact of global warming on Adélie penguins (Ainley was also a consultant for this research):

> In recent decades, Adélie Penguins on the Antarctic continent and on the Antarctic Peninsula have seen very different fortunes. In the northwestern coast of the Antarctic Peninsula, where warming has been dramatic, populations of Adélie Penguins have dropped by 65% over the past 25 years. Temperatures here have risen well above freezing for much of the year. There is less sea ice than before. Antarctic krill and silverfish—Adélie Penguins' primary food source during summer—have been decreasing. Warmer temperatures also have allowed the atmosphere to hold more moisture, thus bringing more snow and reducing the land area on which Adélie Penguins can breed.[18]

But the same study also reported:

> In contrast, along the east coast of the [Antarctic] Peninsula, and on the coast of the Antarctic continent, populations of Adélie Penguins have

> been growing. Here, stronger winds have sustained larger stretches of open water near to the colonies. As a result, it has been easier for the Adélie Penguins to access food which has, until now, been in areas with a lot of sea ice.[19]

Similarly, a 2008 FAQ issued by the British Antarctic Survey also illustrated the variability in the scientific literature pertaining to the impact of global warming on Antarctica's penguins:

> Are penguin populations really declining? Some species are but others are not—it depends on where you look. Long-term monitoring of penguin populations on subantarctic islands reveals a complex picture. In the last 30 years, populations of Adélie penguins on the South Orkney islands have fluctuated, while chinstrap populations have decreased significantly and gentoo [penguin] numbers have risen. On South Georgia, the population of macaroni penguins has declined from 2.5 million breeding pairs in the 1970s to less than 1 million today. King penguins have increased from a few hundred in the 1920s to over 450,000 today. Further south, emperor penguins, which breed on sea ice surrounding continental Antarctica, have also experienced a decline during recent decades—by up to 50% in places.[20]

To illustrate the variability issue further, note that the 2008 British Antarctic Survey reported that "King penguins have increased from a few hundred in the 1920s to over 450,000 today." However, the authors of a study published in the *Proceeding of the National Academy of Sciences* in 2008—the same year in which the British Antarctic Survey FAQ was issued—reported that "our findings suggest that king penguin populations are at heavy risk under the current global warming predictions."[21]

Overall, however, a close reading of the relevant literature shows substantial evidence of penguin vulnerability to global warming beyond the evident levels of regional variability and simple head counts.[22] Furthermore, Lomborg's sidebar "Penguins in Danger?" provided no evidence that environmentalists had exaggerated the danger to penguins from global warming, while ignoring substantial evidence of the harmful impacts of global warming on Emperor penguins and other penguins as reported in many studies.

eight
ON ANTARCTICA AND THE LARSEN-B ICE SHELF

Lomborg's main points about Antarctica in *Cool It* are that the Antarctic continent has cooled in the past several decades, that the dramatic warming of the Antarctic Peninsula is not unprecedented within the past several thousand years, and that Antarctica's contribution to sea-level rise this century will be minimal.[1]

To support his assertion that the Antarctic continent has cooled, Lomborg cited three sources, none of which had been published in a peer-reviewed journal as of *Cool It's* publication date.[2] Lomborg's first source was a draft paper, "A Synthesis of Antarctic Temperatures," that, as the draft noted, was submitted for publication in 2005. Though the paper was published (Chapman and Walsh, 2007) too late for Lomborg to cite in *Cool It*, I will refer to both its unpublished and published versions.

Contrary to what Lomborg reported—that the Antarctic continent has cooled—the published study by Chapman and Walsh reported that much of the Antarctic continent had slightly warmed for much of the recent past, though the authors reported they could detect no significant overall trend: "Trends calculated for the 1958–2002 period suggest modest warming over much of the 60°–90°S domain. All seasons show warming. . . . Because of the large interannual variability of temperatures over the continental Antarctic, most of the continental trends are not statistically significant."[3]

The published draft of the Chapman and Walsh paper thus did not say definitively that the Antarctic continent had cooled; if anything, it reported that the continent had experienced a modest overall warming trend from 1958 to 2002. In addition, the unpublished draft—the one that Lomborg did cite—also reported a "modest warming" on the continent: "Trends calculated for the 1958–2002 period show modest warming over much of 50–90°S with maximum warming over the Antarctic Peninsula."[4]

Furthermore, the unpublished draft appeared to link a detectable warming trend in Antarctica to global warming—enough so that it seemed to undermine Lomborg's assertion that continental Antarctica has cooled and was thus not susceptible to the effects of global warming: "Composite (11-model) GCM-simulations for 1958–2002 with forcing from historic greenhouse gas concentrations show warming patterns and magnitudes similar to the corresponding observed trends.[5]" Overall, neither the published nor the unpublished draft of the Chapman and Walsh study supported Lomborg's claim that the Antarctic continent had undergone a definable cooling; in fact, much of the reported data was to the contrary.

The second source that Lomborg cited to support his claim that the Antarctic continent has cooled was a paper posted by a scientist on the Internet; it did not appear to be a published peer-reviewed study.[6] And the third source could not be located using the Lomborg-provided URL.[7]

Two studies, led by Andrew Monaghan of Ohio State University and published soon after the 2007 publication date of *Cool It*, also did not support Lomborg's claim. One, published in 2008 in the *Bulletin of the American Meteorological Society*, reported "statistically insignificant seasonal and annual near-surface temperature changes over continental Antarctica from the late 1950s through 2000."[8] The second, published in the *Journal of Geophysical Research* in 2008, reported "widespread but statistically insignificant warming over Antarctica from 1992–2005."[9] Indeed, after these reports of statistically insignificant temperature trends, a study published in *Nature Geoscience* (Gillett et al., 2008) reported that Antarctica had significantly warmed, and that the warming was caused by human-induced climate change: "We find that the observed changes in Arctic and Antarctic temperatures are not consistent with internal climate variability or natural climate drivers alone, and are directly attributable to human influence. Our results demonstrate that human activities have already caused significant warming in both polar regions, with likely impacts on polar biology, indigenous communities, ice-sheet mass balance and global sea level."[10]

Another paper (Schneider and Steig, 2008), published in the *Proceedings of the National Academy of Sciences*, reconstructed temperatures over West Antarctica using ice cores, and reported that statistically significant warming had occurred over the Western Antarctic ice sheet during the twentieth century, "underscoring the sensitivity of West Antarctica's climate, and potentially its ice sheet, to large-scale changes in the global climate."[11] A short time later, a major study (Steig et al., 2009), published in *Nature*, reported: "We show that significant warming extends well beyond the Antarctic Peninsula to cover most of West Antarctica, an area of warming much larger than previously reported. West Antarctic warming exceeds 0.1°C per decade over the past 50 years, and is strongest in winter and spring. Although this is partly offset by autumn cooling in East Antarctica, the continent-wide average near-surface temperature trend is positive."[12]

While Lomborg argued, without citing peer-reviewed studies, that Antarctica had cooled, it is clear that the peer-reviewed studies issued after *Cool It*'s 2007 publication date do not support his claim, and that the most recently published studies (2008 and 2009) reported that the Antarctic continent on average, especially West Antarctica, had undergone a significant warming in recent decades.

While misstating temperature trends on the Antarctic continent, Lomborg acknowledged that "the Antarctic Peninsula has warmed dramatically—more than 3.6°F since the 1960s, several times the rate of global warming."[13] However, while seeking to sever the connection between the warming of the Antarctic Peninsula—dramatically illustrated in 2002 by the breakup of the Larsen-B ice shelf along the northeastern coast of the peninsula—and global warming, Lomborg sought to discredit Al Gore's comments about the Larsen-B ice shelf in *An Inconvenient Truth*: "In his film, Al Gore shows us how the ice is rapidly melting and how the less-than-poetically named Larsen B ice shelf dissolved within thirty-five days in early 2002. The significance of this breakup relies on us believing that Larsen B has been intact since time immemorial, so that it now portends dramatically higher sea levels. But this is wrong."[14]

In this passage Lomborg refers to Gore's film, but the reference cites Gore's book.[15] Indeed, Lomborg cites two pages (182–83) that feature four satellite photographs of the Larsen-B ice-shelf breakup. The first photo shows an intact Larsen-B ice shelf still attached to the eastern shore of the Antarctic Peninsula. In the accompanying text, Gore provided the crucial commentary explaining the significance of the Larsen-B breakup:

> The Larsen-B ice shelf, as photographed below, was about 150 miles long and 30 miles wide. When you look at the black pools on top of it, it seems as if you're looking through the ice to the ocean beneath. But that's an illusion. Actually, those are pools of melting water collecting on top of the shelf.
>
> Scientists thought this ice shelf would be stable for at least another century—even with global warming. [Gore then describes the Larsen-B breakup.] Scientists were absolutely astonished. They couldn't figure out how in the world this had happened so rapidly. So they went back to assess why their estimates were off.
>
> They found they had made an incorrect assumption about those melting pools of water on top of the ice mass. They had thought that the meltwater sank back into the ice and refroze. Instead, as they now know, the water keeps sinking straight down and makes the ice mass look like Swiss cheese.[16]

Thus, as Gore points out, the significance of the breakup of the Larsen-B ice shelf is that it prompted a shift in our understanding of ice dynamics—similar to the change in ice dynamics with respect to the Greenland ice sheet—which Lomborg essentially ignored in this instance and throughout *Cool It*. For Lomborg, however, the significance of the Larsen-B breakup was elsewhere. For comparative purpose—which Lomborg repeatedly invited between himself and Gore by regularly invoking Gore's *An Inconvenient Truth*—let's look briefly at two perspectives (Gore v. Lomborg) on the breakup of the Larsen-B ice shelf.

In addition to pointing out that surface meltwater had penetrated and fractured the ice shelf, Gore wrote of the Larsen-B breakup:

"Once the sea-based ice shelf was gone, the land-based ice behind it that was being held back began to shift and fall into the sea. This, too, was unexpected and carries important implications because ice—whether in the form of a mountain glacier or a land-based ice shelf in Antarctica or Greenland—raises the sea level when it melts or falls into the sea. This is one of the reasons sea levels have been rising worldwide, and will continue to go up if global warming is not quickly checked."[17] Gore thus briefly described the significance of the ice dynamics involved in the breakup of the sea-based Larsen-B ice shelf, the potential implications of that discovery on the land-based ice sheets of Greenland and Antarctica, and the ultimate potential impact on sea levels.

Gore regrettably provided no footnotes to support his summary of the significance of the Larsen-B breakup.[18] Nevertheless, there is much scientific evidence to support Gore's synopsis of the implications of Larsen-B as it relates to global warming and sea-level rise. In February 1998, researchers from the British Antarctic Survey published a study in *Nature*, which noted the possibility of a breakup of the Larsen-B ice shelf:

> The breakup of ice shelves has been widely regarded as an indicator of climate change, with observations around the Antarctic Peninsula having shown a pattern of gradual retreat, associated with regional atmospheric warming and increased summer melt and fracturing processes. The rapid collapse of the northernmost section of the Larsen Ice Shelf (Larsen-A) over a few days in January 1995 indicated that, after retreat beyond a critical limit, ice shelves may disintegrate rapidly. . . . Larsen B at present exhibits a stable pattern, but if the ice front were to retreat by a further few kilometers, it too is likely to enter an irreversible retreat phase.[19]

Only one month later, in March 1998, and writing for the U.S. National Snow and Ice Data Center at the University of Colorado, glaciologist Ted Scambos warned of a near-term collapse of Larsen-B: "The Larsen-B appears to have begun the process of breakup, receding past its historical minimum extent, and past the point where recent modeling suggests it can maintain a stable ice front. The breakup appears closely associated with the areas over which melt-ponding is observed during warmer summer sessions."[20]

In September 2001, five months before the breakup of Larsen-B, researchers at the British Antarctic Survey published a report in *Science*, which noted a dramatic warming of the Antarctic Peninsula, with a "99% likelihood that the recent warming is exceptional compared with any part of the 500-year period recorded in the longest of these [ice-core] records." Though the researchers, led by David Vaughan, declined to attribute the warming of the Antarctic Peninsula definitively to "anthropogenic greenhouse gases" without "offering a mechanism," they nevertheless reported: "The recent rapid regional warming in the Antarctic Peninsula is thus exceptional over several centuries and probably unmatched for 1900 years." They also wrote: "Further evidence [of the unusual warming of the Antarctic Peninsula] comes from the retreat of ice shelves, long predicted to result from warming in the Antarctic Peninsula. Rapid regional warming has led to the loss of seven ice shelves during the past 50 years."[21]

Days after the Larsen-B ice shelf had disintegrated, the BBC reported in March 2002 that scientists were surprised by the rapid course of the collapse:

> UK scientists say the Larsen-B shelf on the eastern side of the Antarctic Peninsula has fragmented into small icebergs. Researchers from the British Antarctic Survey (BAS) predicted in 1998 that several ice shelves around the peninsula were doomed because of rising temperatures in the region—but the speed with which the Larsen-B has gone has shocked them. "We knew what was left would collapse eventually, but the speed of it is staggering," said Dr David Vaughan, a glaciologist at BAS in Cambridge. "[It is hard] to believe that 500 billion tonnes of ice sheet has disintegrated in less than a month."[22]

The same BBC article quoted the U.S.-based National Snow and Ice Data Center about the Larsen-B breakup: "This is the largest single event in a series of retreats by ice shelves in the peninsula over the last 30 years. The retreats are attributed to a strong climate warming in the region." These studies and news reports support Gore's description that global warming and meltwater were involved in the Larsen-

B breakup, and that scientists were surprised by the speed of the breakup.

Other studies, summarized below, show that Gore was also correct to say, in the context of Larsen-B, that "once the sea-based ice shelf was gone, the land-based ice sheet behind it that was being held back began to shift and fall into the sea."[23] A year after the disintegration of the Larsen-B ice shelf, *Science* published a study in March 2003 titled, "Glacier Surge After Ice Shelf Collapse." The authors began their report by noting, "The possibility that the West Antarctic Ice Sheet will collapse as a consequence of ice shelf disintegration has been debated for many years. This matter is of concern because such an event would imply a sudden increase in sea level." They concluded: "The evidence presented here unambiguously shows that five of the six major tributaries that formerly nourished the disintegrated portions of LIS [Larsen Ice Shelf] have recently experienced important dynamic perturbations. This includes not only the detected acceleration and retreat [of glaciers] but also an active surging."[24]

Nature News, covering the same *Science* study, similarly reported the implications of the Larsen-B breakup:

> The new findings "raise the likelihood that rapid sea-level rise could be initiated by climate warming," says ice-shelf researcher Ted Scambos at the National Snow and Ice Data Centre in Boulder, Colorado. But there is not enough ice on the Antarctic Peninsula, he points out, to change the world's sea level.
>
> Researchers' biggest fear is that the West Antarctic Ice Sheet, currently hemmed in by the massive Ronne and Ross ice shelves, could go the same way. That would raise sea levels by six or seven metres. "It wouldn't take a very large temperature rise to disintegrate these ice shelves," says [Pedro] Skvarca.[25]

(There is increasing evidence that these comments have validity, and that the impacts of warming that had been confined mainly to the northern Antarctic Peninsula are migrating southward toward West Antarctica. For example, a portion of the Wilkins Ice Shelf, which lies farther south than the Larsen-B area on the Antarctic Peninsula, collapsed in March 2008.[26] This is in addition to the study cited above

[Schneider and Steig, 2008], which reported statistically significant warming over the Western Antarctic ice sheet during the twentieth century.)[27]

A year later, in September 2004, a study published in *Geophysical Research Letters* reported: "Ice velocities derived from five Landsat 7 images between January 2000 and February 2003 show a two-to six-fold increase in centerline speed of four glaciers flowing into the now-collapsed section of the Larsen-B Ice Shelf." The study reported elevation losses within a five- to six-month period of glaciers on the Antarctic Peninsula located along the embayment behind where the Larsen-B ice shelf used to be, including an elevation loss of about thirty-eight meters from the peninsula's Hektoria Glacier. The authors of the study concluded that "both summer melt percolation and changes in the stress field due to [ice] shelf removal play a major role in glacier dynamics."[28]

Thus far, it seems that Gore's analysis of the implications of the Larsen-B breakup—that "once the sea-based ice shelf was gone, the land-based ice behind it that was being held back began to shift and fall into the sea"[29]—accurately reflected the major studies that were issued prior to the publication of *An Inconvenient Truth*. Gore's analysis is also consistent with authoritative commentary (in addition to published peer-reviewed studies) that was issued before and after the publication of his book. For example, in December 2003 the U.S. National Snow and Ice Data Center reported on its Web site, "Scientists at NSIDC have found that glaciers around the area of the Larsen-B Ice Shelf accelerated immediately after it collapsed early in 2002, and are still speeding up. The findings, presented at the AGU [American Geophysical Union] Fall 2003 Meeting in San Francisco, support earlier hypotheses that the ice shelf acted as a barrier, slowing the glaciers as they pushed up against the ice shelf, and that removing the barrier would cause the glaciers to speed up. This finding is significant, because it provides a smaller scale preview of what could occur if larger ice shelves—such as the Ross Ice Shelf [in West Antarctica]—were to collapse."[30]

Similarly, a year after *An Inconvenient Truth* was published, NASA's

Earth System Science Data and Services division summarized the effects of the disintegration of Larsen-B: "After the 2002 Larsen-B ice shelf collapse, glaciers in the embayment emptied into the ocean. Without the buffer of an ice shelf, the glacier's flow accelerated, stretching and thinning the ice."[31]

With this analysis of Gore's discussion about the collapse of the Larsen-B ice shelf on the Antarctic Peninsula, we can now turn to Lomborg's views in *Cool It*. After noting that the Antarctic Peninsula has warmed "more than 3.6°F since the 1960s," and that the "less than poetically named Larsen-B ice shelf dissolved within thirty-five days in early 2002," Lomborg wrote: "The significance of this breakup relies on us believing that Larsen B has been intact since time immemorial, so that it now portends dramatically higher sea levels. But this is wrong."[32] As a prelude to his argument that the warming Antarctic Peninsula and the Larsen-B ice-shelf breakup are not due to human-induced global warming, Lomborg is arguing here that the Larsen-B ice shelf had not been intact "since time immemorial," that its breakup is thus not unprecedented, and that the breakup was due to natural climate fluctuations and does not portend a potentially catastrophic rise in sea levels.

Lomborg did not define an actual time period for "time immemorial," so let's assume that "time immemorial" is analogous to the Holocene geological period, which dates to roughly the past eleven thousand years. Thus, Lomborg is arguing (per our assumption) that the Larsen-B breakup is not unprecedented within this time frame, and that its breakup in 2002 did not have to be interpreted as reflecting any impact of human-induced global warming, or as pointing to any risk of a significant increase in sea levels.[33]

How did Lomborg support these assertions? Oddly, he referenced his key assertion—that "the significance of this breakup relies on us believing that Larsen B has been intact since time immemorial," and that it "is wrong" that the Larsen-B breakup "portends dramatically higher sea levels"—to pages 186–87 of *An Inconvenient Truth*, which show a two-page photograph of "High tide in Funafuti, Tuvalu, Polynesia." The entirety of Lomborg's documentation reads: "Right after Gore's discussion of the Larsen-B breakup, he shows us a large picture of high tides washing in over Tuvalu."[34] Clearly, this reference does nothing to support Lomborg's assertions.

Lomborg then wrote, "Studies show that in the middle of our present interglacial age *the Larsen area* saw 'widespread ice shelf breakup'" (emphasis added).[35] To support this statement Lomborg cited a scientific paper published in 2006 in *Quaternary Science Reviews* (hereinafter referred to as the Pudsey paper, after its lead author, Carol J. Pudsey).[36] The Pudsey paper refers to a "widespread ice-shelf breakup" in the mid-Holocene in "the northern Larsen area," but not in the area of the Larsen-B ice shelf, which lies in the central part of the larger Larsen area. This is a significant problem since "the Larsen area"—as Lomborg wrote—consists of three distinct ice shelves: the northern Larsen-A ice shelf, which disintegrated in 1995; the southern Larsen-C ice shelf, which is still intact; and the central Larsen-B ice shelf, which lies between the Larsen-A and Larsen-C areas, and which disintegrated in 2002. Lomborg muddled these key distinctions by implying that "the Larsen area" is synonymous and interchangeable with the Larsen-B area. Also, the stated focus of the Pudsey paper was to examine the "continental shelf sediments in the northern Larsen area" following the 1995 Larsen-A ice-shelf breakup.[37] It was therefore not a study of the Larsen-B ice shelf or the Larsen-B area. Thus, while supposedly demonstrating that the 2002 Larsen-B ice shelf breakup had a precedent during the Holocene, Lomborg dropped the specific reference to Larsen-B, inserted the broader "Larsen area" reference, and cited a study about the Larsen-A area as if it supported his claims about the Larsen-B area. In other words, Lomborg's source for his claim that the Larsen-B ice shelf breakup in 2002 had a precedent during the Holocene—the Pudsey paper—provided no such evidence.

Similarly, still in the context of arguing that the Larsen-B breakup was not unprecedented, and continuing to reference "the Larsen area" as if it were interchangeable with the Larsen-B area, Lomborg wrote: "It is likely that *the Larsen area* was open water from perhaps six thousand to two thousand years ago" (emphasis added).[38] Lomborg cited two sources here, the first one to the Pudsey paper.[39] However, the only mention of "open water" pertaining to this source is in a paragraph about the Larsen-A area, which is north of the Larsen-B area, and the Prince Gustav Channel, which lies north of the Larsen-A area, and which is thus even farther from the Larsen-B area. Once again, while referring generically to "the Larsen area," Lomborg referenced a page in the Pudsey paper that does not refer to the Larsen-B area.

Likewise, the second source that Lomborg cited to support his claim that "the Larsen area" was open water six to two thousand years ago—a 2001 study published in *Science*—also is not focused on the Larsen-B area.[40] In fact, the only mention of "open water" in the *Science* study states (in referring to a contemporary situation): "Adélie penguins, which require access to winter pack ice, are declining around Faraday, whereas chinstrap penguins, which usually require open water, are increasing."[41] This sentence has nothing to do with whether the Larsen-B area was "open water" "from six thousand to two thousand years ago," as Lomborg wrote. Indeed, on a map of the Antarctic Peninsula that is featured in the *Science* analysis, the only two locations identified along the entire northeastern side of the peninsula (excluding the small islands at the extreme tip of the peninsula) are "Prince Gustav Channel" and "Larsen Ice Shelf-A."[42] Neither of these sources that Lomborg cited supported—indeed, even addressed—Lomborg's contention regarding the breakup of the Larsen-B ice shelf.

At the same time, Lomborg ignored a major study that focused specifically on the Larsen-B ice shelf. This study, titled "Stability of the Larsen-B Ice Shelf on the Antarctic Peninsula During the Holocene Epoch," was published in *Nature* in 2005, well before the 2007 publication date of *Cool It*. Led by Hamilton College geologist Eugene Domack, this study reported: "Here we use records of diatoms, detrital material and geochemical parameters from six marine sediment cores in the vicinity of the Larsen ice shelf to demonstrate that the recent collapse of the Larsen B ice shelf is unprecedented during the Holocene."[43] The authors of the study did note a "thinning throughout the Holocene" of the Larsen-B ice shelf, and concluded: "We suggest that the recent prolonged period of warming in the Antarctic Peninsula region, in combination with the long-term thinning, has led to the collapse of the ice shelf."[44] These findings, which actually pertain specifically to the Larsen-B ice shelf, are certainly closer to Gore's analysis that the collapse of the Larsen-B was a unique environmental event with serious implications for global warming and sea-level rise, than to Lomborg's, which denied any such implications of the Larsen-B collapse.

The conclusion that global warming played a major role, but perhaps not the only role, in the Larsen-B breakup was supported more recently in a study published in the *Journal of Glaciology* in 2008.[45] An article in *ScienceDaily* about the study reported:

> "Ice shelf collapse is not as simple as we first thought," said Professor [N. F.] Glasser, lead author of the paper. "Because large amounts of meltwater appeared on the ice shelf just before it collapsed, we had always assumed that air temperature increases were to blame. But our new study shows that ice-shelf break up is not controlled simply by climate. A number of other atmospheric, oceanic and glaciological factors are involved. For example, the location and spacing of fractures on the ice shelf such as crevasses and rifts are very important too because they determine how strong or weak the ice shelf is."[46]

ScienceDaily noted that "Professor Glasser acknowledges that global warming had a major part to play in the collapse, but emphasizes that it is only one in a number of contributory factors."[47]

In his description of the elements behind the Larsen-B ice-shelf collapse, it seems that Gore accurately reflected the state of knowledge of the Larsen-B breakup contemporaneous to 2006 (the publication date of *An Inconvenient Truth*), including with respect to the meltwater and crevasses Gore featured in his book and film, in addition to the major role played by global warming.[48]

Still alluding to but not specifically referring to Larsen-B, and continuing his argument that the 2002 collapse of the Larsen-B ice shelf had a precedent during the Holocene, Lomborg wrote: "The maximal ice shelf dates only from the Little Ice Age a couple of hundred years ago, and much of what has subsequently collapsed is of that vintage."[49] A source for this assertion is not readily evident, though the only possible sources Lomborg could be referencing here are the 2006 Pudsey study published in *Quaternary Science Reviews*,[50] and the 2001 *Science* report.[51] Although the Pudsey paper is focused only on "the northern Larsen area" and "Prince Gustav Channel," it reports that "the maximum ice shelf limit may date only from the Little Ice Age."[52] Since this language is nearly identical to the language used by Lomborg writing in the context of his claims about the Larsen-B area—that "the maximal ice shelf dates only from the Little Ice Age"—one might conclude that this reference from the Pudsey paper is Lomborg's source for this claim. But as stated previously, the Pudsey paper was not a study of the Larsen-B area.

Lomborg continued: "Moreover, the breakup of the ice shelf did not cause the sea level to rise, because it was already floating."[53] Given the context of Lomborg's narrative, and previous references to Al Gore and the Larsen-B ice shelf, the appearance here is that Lomborg is responding to an allegation from someone, presumably Gore, that the Larsen-B collapse itself caused sea levels to rise. But Gore—the only person mentioned on this page in *Cool It*, or in the immediately preceding or following pages—never said that the Larsen-B ice shelf collapse by itself would lead to sea level rise. In fact, Gore painstakingly explained how the Larsen-B collapse, and the collapse of other sea-based ice-shelves, could contribute indirectly to sea-level rise, since ice shelves function as a barrier between land-based ice and the sea, without which the land-based ice could slide more quickly into the sea, and thus increase sea levels.[54]

Following his discussion of the collapse of the Larsen-B ice shelf, Lomborg proceeded to make his next erroneous claim—that the Antarctica Peninsula, to which the Larsen-B ice shelf was attached until early 2002, is contributing to a net *decrease* in sea level. Lomborg wrote: "While it [the Larsen-B collapse] probably led to ice shelves flowing more quickly into the sea and [land-based] glaciers retreating [melting] at a faster pace, the story [unidentified, but presumably

Gore's 'story' about the Larsen-B collapse] left out one important fact. The precipitation on the Antarctic Peninsula is increasing, probably due to climate change, and this likely outweighs the melting. That is, despite the spectacular pictures of Larsen B, the Antarctic Peninsula is probably participating in an overall *lowering* of sea levels" (emphasis in original).[55]

In this passage, Lomborg again misapplied the relevance of the collapse of the Larsen-B ice shelf. The Antarctic Peninsula, from which the Larsen-B ice shelf broke, and which makes up roughly 4 percent of the Antarctic continent, is not big enough by itself (as noted above) to significantly affect sea-level rise. The sea-level relevance of the Larsen-B ice shelf breakup is the fact that what happened to Larsen-B on the Antarctic Peninsula could happen to other ice shelves along the coasts of continental Antarctica. These bigger ice shelves essentially stand between the huge ice sheet of the continent and the open sea. Thus, the major sea-level implications of the collapse of the Larsen-B ice shelf do not pertain directly to the Antarctic Peninsula itself, but to the ultimate fate of the ice shelves buttressing the Antarctic ice sheet, and to whether the breakup of the Larsen-B ice shelf along the Antarctic Peninsula portends a breakup of the other ice shelves along the outer rim of Antarctica, especially West Antarctica, which in turn would threaten a breakup of the Greenland-size West Antarctic ice sheet. Thus, going back to Lomborg's "one important fact" that someone "left out" of the Larsen-B "story"—that "despite the spectacular pictures of Larsen-B, the Antarctic Peninsula is probably participating in an overall *lowering* of sea levels"—that "fact" would not be all that relevant, given the larger implications for sea-level rise of the Larsen-B collapse.

Nor is this "important fact" arguably even a fact, at least not in the long run. When Lomborg wrote that "precipitation on the Antarctic Peninsula is increasing, probably due to climate change, and this likely outweighs the melting,"[56] he cited four sources,[57] one of which (Morris and Mulvaney, 2004) noted that the peninsula would contribute a small increase in sea levels if recent warm conditions persist. And a study published a few months prior to the publication of *Cool It* reported that the Antarctic Peninsula contributed a small net *increase* in sea levels from 1993 to 2003 (Pritchard and Vaughan, 2007).[58]

Overall, Lomborg provided no significant evidence in the "Rising Sea Levels" section of *Cool It* to support any of his claims: the Antarctic continent had cooled in recent decades; there is a precedent for the collapse of the Larsen-B ice shelf during the Holocene; the Larsen-B area was "open water" during the Holocene; and the Larsen-B breakup signaled no threat of higher sea levels due to global warming.

nine
ON HURRICANES AND EXTREME WEATHER EVENTS

In addition to arguing that environmentalists have exaggerated the threat of global warming to polar bears, penguins, and sea-level rise, Lomborg also argued that environmentalists have exaggerated the link between global warming and extreme weather events, including hurricanes, heat waves, heavy precipitation events, and drought. Lomborg argued that few such events will increase due to global warming, or at least to the degree that the environmentalists claim. This point of view thus explains the title of this section of *Cool It*, "Extreme Weather, Extreme Hype."

Lomborg began by complaining that the Natural Resources Defense Council, Friends of the Earth, and Greenpeace inappropriately link extreme weather events, such as hurricanes, to global warming. Their remedy, Lomborg complained, "is invariably CO_2 cuts and adoption of Kyoto."[1] Lomborg's objective is to show that the claims of these organizations are inaccurate or exaggerated with respect to the link between global warming and extreme weather events, and that there is no need on this count to reduce greenhouse emissions.

Al Gore is on Lomborg's list of exaggerators in this respect as well. To illustrate how he claims Gore exaggerated the connection between global warming and hurricanes in *An Inconvenient Truth*, including the impact of Hurricane Katrina on New Orleans in 2005, Lomborg wrote: "Al Gore spends twenty-six pages showing pictures of the suffering in New Orleans and names every single hurricane in 2005."[2] Lomborg doesn't reference this claim, so we don't know exactly to which pages in *An Inconvenient Truth* he was referring. Upon turning through every page of Gore's book, however, I counted four pages of pictures—not twenty-six—showing the suffering in New Orleans from Hurricane Katrina.[3] The entire section on Hurricane Katrina in Gore's book (including pages with no pictures of New

Orleans) consists of eight pages.[4] Thus, here is another inaccurate allegation by Lomborg of an environmentalist's exaggeration.

Lomborg next criticized Robert F. Kennedy Jr. for linking fossil-fuel dependence, global warming, the destructiveness of Hurricane Katrina, and the tragedy of New Orleans: "Robert F. Kennedy Jr., when looking at the New Orleans tragedy, blamed it on the United States 'derailing the Kyoto Protocol' and said that 'now we are all learning what it's like to reap the whirlwind of fossil fuel dependence.' "[5] Lomborg, however, did not mention the basis on which Kennedy invoked these links — a study published in *Nature* on August 4, 2005, that is, three weeks before Katrina hit New Orleans on August 26. The *Nature* study, titled "Increasing Destructiveness of Tropical Cyclones Over the Past 30 Years," by Kerry Emanuel of the Massachusetts Institute of Technology, reported "longer storm lifetimes and greater storm intensities" due to global warming. Emanuel reported: "I find that the record of net hurricane power dissipation is highly correlated with tropical sea surface temperature, reflecting well-documented climate signals, including multi-decadal oscillations in the North Atlantic and North Pacific, and global warming." Hurricane "power dissipation" is the energy absorbed by a hurricane from the warm waters below it, which power the force and speed of its winds. Emanuel found a "near doubling of power dissipation over the period of record" (from the mid-1970s) in his study. He concluded: "My results suggest that future warming may lead to an upward trend in tropical cyclone destructive potential, and — taking into account an increasing coastal population — a substantial increase in hurricane-related losses in the twenty-first century."[6]

Kennedy and Gore both cited this study, but Lomborg does not mention it. Kennedy wrote: "This month, a study published in the journal Nature by a renowned M.I.T. climatologist linked the increasing prevalence of destructive hurricanes to human-induced global warming."[7] Likewise Gore wrote, in the first page of his section on Hurricane Katrina in *An Inconvenient Truth*, "On July 31, 2005, less than a month before Hurricane Katrina hit the United

States, a major study from M.I.T. supported the scientific consensus that global warming is making hurricanes more powerful and more destructive."[8] Not only did Lomborg neglect to cite the M.I.T./*Nature* study in the context of his criticism of Gore and Kennedy, but he also didn't cite it within the framework of his own analysis of hurricanes and global warming in *Cool It*.[9]

Furthermore, as Gore indicated, the M.I.T./*Nature* study reflected the scientific consensus on extreme weather events and global warming, including as reported in the 2001 IPCC assessment: "Extreme events are a major source of current climate impacts, and changes in extreme events are expected to dominate the impacts of climate change;"[10] "The frequency and magnitude of many extreme climate events increase even with a small temperature increase and will become greater at higher temperatures";[11] "Extreme events include, for example, floods, soil moisture deficits, tropical cyclones [hurricanes], storms, high temperatures, and fires."[12] And from the "Summary for Policymakers," typically the most widely read chapter of the IPCC assessments, the 2001 IPCC assessment report stated: "Models project that increasing atmospheric concentrations of greenhouse gases result in changes in frequency, intensity, and duration of extreme events, such as more hot days, heat waves, heavy precipitation events, and fewer cold days. Many of these projected changes would lead to increased risks of floods and droughts in many regions, and predominantly adverse impacts on ecological systems, socio-economic sectors, and human health. High resolution modeling studies suggest that peak wind and precipitation intensity of tropical cyclones are likely to increase over some areas."[13] Using the 2001 IPCC assessment report as the scientific benchmark, it seems fair to conclude that it reflected a "scientific consensus," as Gore wrote, that the intensity of hurricanes has increased due to global warming, and will increase further as the Earth and its oceans grow warmer.

What's more, the 2007 IPCC assessment report extensively discussed the connection between global warming and extreme weather events, including hurricanes, and Lomborg included none of these findings in *Cool It.* Of the connection the report states:

> People affected by an extreme weather event (e.g., the extremely hot summer in Europe in 2003, or the heavy rainfall in Mumbai, India in July 2005) often ask whether human influences on the climate are responsible for the event. A wide range of extreme weather events is expected in most regions even with an unchanging climate, so it is difficult to attribute any individual event to a change in climate. . . . However, simple statistical reasoning indicates that substantial changes in the frequency of extreme events (and in the maximum feasible extreme, e.g., the maximum possible 24-hour rainfall at a specific location) can result from a relatively small shift of the distribution of a weather or climate variable.[14]

In a full-page, two-color sidebar analysis that asked, "Has There Been a Change in Extreme Events like Heat Waves, Droughts, Floods and Hurricanes?" the 2007 IPCC assessment observed:

> Globally, estimates of the potential destructiveness of hurricanes show a substantial upward trend since the mid-1970s, with a trend towards longer storm duration and greater storm intensity, and the activity is strongly correlated with tropical sea surface temperature. These relationships have been reinforced by findings of a large increase in numbers and proportion of strong hurricanes globally since 1970 even as total numbers of cyclones and cyclone days decreased slightly in most basins. Specifically, the number of category 4 and 5 hurricanes increased by about 75% since 1970. The largest increases were in the North Pacific, Indian and Southwest Pacific Oceans. However, numbers of hurricanes in the North Atlantic have also been above normal in 9 of the last 11 years, culminating in the record-breaking 2005 season.[15]

Two pages later, the IPCC 2007 assessment prominently presented another sidebar titled, "Recent Extreme Events." This three-page section listed "some recent notable extreme climate events,"[16] including the 2005 hurricane season (recall that Lomborg criticized Gore apparently for even mentioning it):

- Drought in Central and Southwest Asia, 1998–2003
- Drought in Australia, 2002–2003

- Drought in Western North America, 1994–2004
- Floods in Europe, Summer 2002
- Heat Wave in Europe, Summer 2003
- The 2005 Tropical Storm Season in the North Atlantic

About the "2005 Tropical Storm Season in the North Atlantic," the 2007 IPCC assessment report concluded:

> The 2005 North Atlantic hurricane season (1 June to 30 November) was the most active on record by several measures, surpassing the very active season of 2004 and causing an unprecedented level of damage. Even before the peak in the seasonal activity, the seven tropical storms in June and July were the most ever, and hurricane Dennis was the strongest on record in July and the earliest ever fourth-named storm. The record 2005 North Atlantic hurricane season featured the largest number of named storms (28) and is the only time names have ventured into the Greek alphabet. It had the largest number of hurricanes (15) recorded, and is the only time there have been four category 5 storms. These included the most intense Atlantic storm on record, the most intense storm in the Gulf of Mexico, and Katrina.[17]

Gore's 2006 commentary in *An Inconvenient Truth* on the 2005 hurricane season was consistent with what the IPCC would report in 2007 and, contrary to Lomborg's assertions, was hardly exaggerated in the context of the scientific evidence reviewed above. Gore wrote:

> Hard on the heels of 2004 came the record-breaking summer of 2005. Several hurricanes hit the Caribbean and the Gulf of Mexico early in the season, including Hurricane Dennis and Hurricane Emily, which caused significant damage.
>
> The emerging consensus linking global warming to the increasingly destructive power of hurricanes has been based in part on research showing a significant increase in the number of category 4 and 5 hurricanes.
>
> A separate study predicts that global warming will increase the strength of the average hurricane a full half-step on the well-known five-step scale.
>
> The National Oceanic and Atmospheric Administration summarized some of the basic elements common to these new research studies in the graph shown below. [The graph shows that as water temperature increases, wind velocity and the moisture content of hurricanes increases.] As water temperature goes up, wind velocity goes up, and so does storm moisture condensation.[18]

Although Lomborg had access to the 2007 IPCC assessment report while Gore did not, Gore's analysis of global warming and hurricanes was practically a portent of the 2007 IPCC report; yet Lomborg, who ignored the 2007 IPCC report on hurricanes and global warming, accused Gore of exaggerating the connection between global warming and hurricanes. Lomborg even wrote that "these statements [about hurricanes from Gore, Kennedy, and Ross Gelbspan] resemble the exaggerated stories of the polar bears."[19]

Lomborg also accused environmentalists of telling exaggerated "stories"—a much-used word in *Cool It* that conveniently muddles the precise subject and nature of Lomborg's complaints—about global warming and flooding rivers. Lomborg thus began yet another alleged exposé of a global warming exaggeration, titled "Flooding Rivers."

He began his argument by stating that "the story of river flooding is much the same as what we saw with hurricanes."[20] He acknowledged that "unusually severe floods in the 1990s and the early 2000s from St. Louis in the United States to Poland, Germany, France, Switzerland, Spain, and the United Kingdom have garnered renewed attention to the problem of flooding." However, he then wrote, "Very often, those commenting on these floods make an explicit link to climate change."[21] Lomborg identified the advocates of this "explicit link" as British Prime Minister Tony Blair, French President Jacques Chirac, and German Chancellor Gerhard Schröder. However, not only did none of these heads of state make such a link, at least as reported by Lomborg, it is doubtful that they would have, given the inherent scientific uncertainty in definitively linking any *single* extreme weather event to climate change. This is not to say, however, as we shall see below, that the severe floods in Europe in 2002 could not be linked to global warming. Lomborg thus muddled the distinction between citing a *provable* link as opposed to citing the scientific *probability* or *likelihood* of a link in order to argue that there was no link between the flood and global warming.

Lomborg then wrote:

> After the severe flood of Prague and Dresden in 2002, British prime minister Tony Blair, French president Jacques Chirac, and German chancellor Gerhard Schröder all used the flood as a prime example of why we must commit to Kyoto. According to Schröder, this flood showed us that "climate change is no longer a skeptical prognosis, but a bitter reality. This challenge demands decisive action from us," which he identified as a requirement for "all states to ratify the Kyoto Protocol."[22]

Not only did this passage neglect to show that Blair, Chirac, and Schröder made any explicit link between the 2002 European floods and climate change, as Lomborg asserted, it also incorrectly implied that any assumption that the 2002 floods were linked to global warming was invalid.

After noting that German Chancellor Schröder commented that "climate change is no longer a skeptical prognosis, but a bitter reality," Lomborg appeared to offer the following small concession: "And yes, it is true that global warming *eventually* will increase precipitation, especially heavy rains" (emphasis added).[23] Lomborg's point here, presumably, is that global warming "eventually" will increase precipitation but hasn't done so yet. Lomborg cited three sources for this statement: a published study (Groisman et al., 2005) and two sections from the 2007 IPCC assessment. However, neither Groisman nor the IPCC supports Lomborg's claim that increased precipitation due to global warming will only "eventually" happen; instead, these sources reported that increased precipitation due to global warming is *already* happening, and will continue. Groisman wrote: "In the mid-latitudes, there is a widespread increase in the frequency of very heavy precipitation during the past 50 to 100 yr." Groisman linked this recent increase to human-induced global warming: "It was found that both the empirical evidence from the period of instrumental observations and model projections of a greenhouse-enriched atmosphere indicate an increasing probability of intense precipitation events for many extratropical regions including the United States."[24] Likewise, the 2007 IPCC assessment report stated: "Changes in some types of extreme events have already been observed, for example, increases in the frequency and intensity of heatwaves and heavy precipitation events."[25] Thus, Lomborg's own sources reported that the incidence of heavy precipitation events had already increased as a result of human-induced global warming, though Lomborg cited these sources to support his claim that such events would only "eventually" increase.

Lomborg's next sentence also is not supported by his sources. Lomborg wrote: "Models also show that this *will lead* to more flooding" (emphasis added).[26] In the context of his previous claim, Lomborg is saying that as heavy precipitation events "eventually" increase, climate models show that the increase in precipitation "will [*eventually*] lead to more flooding" in the future. But Lomborg's source, a 2002 study published in *Nature* titled, "Increasing Risk of Great Floods in a Changing Climate," did not speak exclusively in the future tense. It reported: "We find that the frequency of great floods increased substantially during the twentieth century. The recent emergence of a statistically significant positive trend in risk of great floods is consistent with results from the climate model, and the model suggests that the trend will continue."[27] These findings would have been difficult for Lomborg to miss, since they were reported in the abstract of the study in *Nature* that Lomborg himself referenced.[28] Thus, despite giving the impression that extreme heavy precipitation and flooding events due to global warming might occur at some point in the future—this to help support his argument that recent floods in Europe were not caused by global warming—Lomborg's sources reported that such events are already occurring, and will likely continue into the future.

Lomborg's next sentence conceded that "there is also some evidence that increased rain is already occurring, although the IPCC has still not been able to link it to global warming."[29] Lomborg's principal source to support this claim is a 2007 IPCC assessment report sidebar titled, "Can Individual Extreme Events Be Explained by Greenhouse Warming?"[30] The sidebar mentions heavy rainfall in the last paragraph, which, on a "likelihood" basis, and contrary to Lomborg's claim, affirms a connection between the recent history of heavy rainfalls and floods to human-induced global warming: "The same likelihood-based approach can be used to examine changes in the frequency of heavy rainfall or floods.

Climate models predict that human influences will cause an increase in many types of extreme events, including extreme rainfall. There is already evidence that, in recent decades, extreme rainfall has increased in some regions, leading to an increase in flooding."[31] Thus, contrary to what Lomborg reported, the 2007 IPCC assessment linked heavy rainfalls and floods to human-induced global warming on the basis of a scientific "likelihood." This is very different from Lomborg's claim that the IPCC "has still not been able to link" these events to global warming.

Lomborg also argued—in the context of the 2002 floods in Europe and further obscuring the link between the floods and global warming—that an increase in heavy precipitation events does not lead to an increase in river flooding. In making this case, Lomborg again invoked former German Chancellor Schröder: "But there are two problems with Schröder's argument. First, the increasing rain does not seem to be translated into increasing flooding in rivers."[32] For one thing, it is absurd to argue, in the context of the 2002 floods in Europe, that an increase in heavy precipitation events would not lead to an increase in river flooding, given that the 2002 floods, including major flooding in the Czech Republic and Germany, were preceded by a week of continuous rain.[33] In addition, Lomborg's argument is not supported by the 2007 IPCC assessment report, which clearly attributed a higher incidence of river flooding to increased rainfall, with negative impacts:

- "Increased precipitation intensity and variability is projected to increase the risk of floods and droughts in many areas."[34]
- "Drought-affected areas will probably increase, and extreme precipitation events, which are likely to increase in frequency and intensity, will augment flood risk."[35]
- "Projected surface warming and shifts in rainfall in most countries of Asia will induce substantial declines in agricultural crop productivity as a consequence of thermal stress and more severe droughts and floods."[36]
- "Increases in rainfall in south-east Brazil, Paraguay, Uruguay, the

Argentine Pampas, and some parts of Bolivia have had impacts on land use and crop yields and have increased flood frequency and intensity."[37]

- "More intense rain and more frequent flash floods during the monsoon [in Asia] would result in a higher proportion of runoff and a reduction in the proportion reaching the groundwater."[38]

Although he wrote that "increasing rain does not seem to be translated into increasing flooding in rivers,"[39] Lomborg did not engage these IPCC findings, which clearly link increased rainfall to an increase in flooding.

Instead of citing the 2007 IPCC assessment report, Lomborg cited two 2005 studies to support his claim that an increase in rain does not lead to an increase in flooding.[40] Thus, Lomborg wrote: "First, the increasing rain does not seem to be translated into increasing flooding in rivers. This holds true in a global sample of almost two hundred rivers, in which twenty-seven did indeed show increasing high flows, but even more (thirty-one) rivers were decreased, and the large majority showed no trend. This also holds true for the smaller number of rivers around the world where we have observations stretching very far back. Why is this?"[41] Without more information, and because Lomborg cited broad global trends that reflect no specific relationships between an increase or decrease in rainfall and a corresponding increase or decrease in flooding events, it is not clear how this passage supports his claim that "increasing rain does not seem to be translated into increasing flooding in rivers."

Furthermore, the fact that some rivers would show an increase in flows while some would show a decrease would not necessarily undermine an increased risk of river flooding because of global warming, since global warming has shifted rainfall patterns worldwide. As a result of this shift, some regions have experienced an increase in precipitation and a higher risk of river flooding, while other regions have experienced a decrease in precipitation and a higher risk of drought. This situation not only was clearly indicated in the brief excerpts above from the 2007 IPCC assessment report, but was also reported in the IPCC's 2007 detailed assessment of the impact of shifting rainfall patterns in Asia, which have led already to an increased frequency in flooding (due to increased rainfall) and drought (due to decreased rainfall):

> Intense Rains and Floods
> Russia: Increase in heavy rains in western Russia and decrease in Siberia; increase in number of days with more than 10 mm rain; 50 to 70% increase in surface runoff in Siberia.
> China: Increasing frequency of extreme rains in western and southern parts including Changjiang river, and decrease in northern regions; more floods in Changjiang river in past decade; more frequent floods in North-East China since 1990s; more intense summer rains in East China; severe flood in 1999; seven-fold increase in frequency of floods since 1950s.
> Japan: Increasing frequency of extreme rains in past 100 years attributed to frontal systems and typhoons; serious flood in 2004 due to heavy rains brought by 10 typhoons; increase in maximum rainfall during 1961 to 2000 based on records from 120 stations.
> South Asia: Serious and recurrent floods in Bangladesh, Nepal and north-east states of India during 2002, 2003 and 2004; a record 944 mm of rainfall in Mumbai, India on 26 to 27 July 2005 led to loss of over 1,000 lives with loss of more than US$250 million; floods in Surat, Barmer and in Srinagar during summer monsoon season of 2006; 17 May 2003 floods in southern province of Sri Lanka were triggered by 730 mm rain.
> South-East Asia: Increased occurrence of extreme rains causing flash floods in Vietnam; landslides and floods in 1990 and 2004 in the Philippines, and floods in Cambodia in 2000.
>
> Droughts
> Russia: Decreasing rain and increasing temperature by over 1°C have caused droughts; 27 major droughts in 20th century have been reported.
> Mongolia: Increase in frequency and intensity of droughts in recent years; droughts in 1999 to 2002 affected 70% of grassland and killed 12 million livestock.
> China: Increase in area affected by drought has exceeded 6.7 Mha since 2000 in Beijing, Hebei Province, Shanxi Province, Inner Mongolia and North China; increase in dust storm affected area.

South Asia: 50% of droughts associated with El Niño; consecutive droughts in 1999 and 2000 in Pakistan and N-W India led to sharp decline in watertables; consecutive droughts between 2000 and 2002 caused crop failures, mass starvation and affected ~11 million people in Orissa; droughts in N-E India during summer monsoon of 2006.

South-East Asia: Droughts normally associated with ENSO years in Myanmar, Laos, Philippines, Indonesia and Vietnam; droughts in 1997 to 98 caused massive crop failures and water shortages and forest fires in various parts of Philippines, Laos and Indonesia.[42]

While Lomborg features North America and Europe in his section on "Flooding Rivers,"[43] he never mentions Asia. Lomborg also ignored Latin America, about which the 2007 IPCC assessment report noted, also with respect to rainfall, droughts, and floods: "Climatic variability and extreme events have been severely affecting the Latin America region over recent years (high confidence). Highly unusual extreme weather events were reported, such as intense Venezuelan rainfall (1999, 2005), flooding in the Argentinean Pampas (2000–2002), Amazon drought (2005), hail storms in Bolivia (2002) and the Great Buenos Aires area (2006), the unprecedented Hurricane Catarina in the South Atlantic (2004) and the record hurricane season of 2005 in the Caribbean Basin."[44] Referring to these events, the IPCC observed: "The occurrence of climate-related disasters increased by 2.4 times between the periods 1970–1999 and 2000–2005 continuing the trend observed during the 1990s."[45]

It thus seems clear—even when a given extreme weather event cannot be said definitively to be an impact of global warming—that the incidence of extreme weather events has increased in recent decades as a result of global warming, including more frequent heavy precipitation events and river flooding.

ten
MALARIA IN VERMONT

Lomborg began his section in *Cool It* called "Malaria in Vermont" by referring to a 2003 World Health Organization report titled, "Climate Change and Human Health: Risks and Response."[1] As Lomborg noted,[2] the WHO estimated that global warming had caused 150,000 excess deaths in 2000.[3] "Green organizations, political parties, and pundits have incessantly repeated this figure," he wrote.[4] As evidence that this figure had been "incessantly" repeated, Lomborg provided just three sources:[5] a 2006 article in the British newspaper the *Guardian*;[6] a report by the British Liberal Democratic Party, which could not be found using the Lomborg-provided URL or the document title;[7] and a 2005 article posted on the Greenpeace U.K. Web site.[8] Neither the newspaper article nor the Greenpeace Web site article provide any analysis about how often the WHO statistic was cited.

Lomborg continued: "Not surprisingly, a headline like 'Climate Change Death Toll Put at 150,000' sells a lot of newspapers."[9] To support this claim, Lomborg cited one source: a Reuters dispatch that was posted on the Common Dreams News Center Web site.[10] However, Reuters is not a newspaper, and access to the Common Dreams Web site is free.[11]

Lomborg then attempted to show how the WHO itself was guilty of exaggeration by arriving at that statistic in the first place. After commenting that the statistic of 150,000 excess fatalities sells newspapers, and for the purpose of exposing this alleged WHO exaggeration, Lomborg invited readers to "take a peek under the hood of this number."[12] However, the 2003 WHO report almost certainly underreported excess deaths by considering only some impacts on human health caused by global warming, while not considering others. The report stated:

> Outcomes To Be Assessed
>
> While a wide variety of disease outcomes is suspected to be associated with climate change, only a few outcomes are addressed in this analysis (Table 7.2). These were selected on the basis of:
>
> - sensitivity to climate variation
> - predicted future importance
> - availability of quantitative global models (or feasibility of constructing them).

In reviewing the strength of the evidence for each of these areas, the WHO report pointed to all papers in the health section of the 2001 IPCC assessment, other wide-ranging reviews of climate change and health, and a systematic review of the scientific literature using relevant Internet search engines (Medline and Web of Science). The report continued:

> Additional likely effects of climate change that could not be quantified at this point include:
>
> - changes in pollution and aeroallergen levels
> - recovery rate of the ozone hole, affecting exposure to UV radiation
> - changes in distribution and transmission of other infectious diseases (particularly other vector-borne diseases and geohelminths [parasitic worms])
> - indirect effects on food production acting through plant pests and diseases
> - drought
> - famine
> - population displacement due to natural disasters, crop failure, water shortages
> - destruction of health infrastructure in natural disasters
> - risk of conflict over natural resources.
>
> Some of these may be included in future assessments as additional quantitative evidence becomes available.[13]

In taking his readers "under the hood" of the WHO report, Lomborg did not mention that the estimate of 150,000 excess deaths in 2000 excluded several health-related impacts of global warming. Lomborg then wrote that "a curious thing happened with cold and heat deaths," as estimated in the WHO report: "While the [WHO] authors spent three pages talking about heat and cold deaths, when they aggregated the numbers they simply *left out* cold and heat deaths, leading to the total death toll of 153,000" (emphasis in original).[14] In actuality, the authors of the report explained that the cold and heat deaths may cancel each other out (words in parentheses in original):

> Climate change is expected to affect the distribution of deaths from the direct physiological effects of exposure to high or low temperatures (i.e. reduced mortality in winter, especially in high latitude countries, but increases in summer mortality, especially in low latitudes). However, the overall global effect on mortality is likely to be more or less neutral. The effect on the total burden of disease has not been estimated, as it is unclear to what extent deaths in heat extremes are simply advancing deaths that would have occurred soon in any case.[15]

Though the WHO painstakingly noted how it quantified the health impacts of specific human risk factors regarding global warming,[16] and while accusing the WHO authors of a "curious" omission of statistics pertaining to cold- and heat-related deaths, Lomborg, in contrast to the rigorous methods of the WHO authors, provided his own "rough estimate" of human mortality in this regard. Lomborg wrote: "If we make a rough estimate of the lives lost and saved by the temperature increase since the 1970s of 0.65°F, we get about 620,000 avoided cold deaths and 130,000 extra heat deaths. This of course dramatically influences the total outcome: instead of 150,000 dying of global warming, there are actually almost 200,000 more people surviving each year."[17]

Lomborg's only reference to document the numbers in this key passage illustrates how his sources often confuse his factual assertions rather than substantiate them. This reference reads, in its entirety, "(CRU, 2006) shows 0.361°C change from the 1961–90 average in 2000; (WHO, WMO, & UNEP, 2003:7) estimates 0.4°C. The estimate comes from a linear extrapolation from (Bosello, Roson, & Tol, 2006), which estimates increases of 1.03°C from today's temperature.

Taking the proportional (0.35 = 0.361/1.03) cold and heat deaths gives the numbers here. It also gives an estimate of all other illnesses of 193,000, which compares fairly well with WHO's 150,000."[18]

By way of explication, the first sentence presumably shows how Lomborg estimated that the global average temperature since 1961 has increased by 0.65°F. (The Fahrenheit conversion for 0.361°C is 0.65°F, and the conversion for 0.4°C is 0.72°F.) Beyond these apparent temperature conversions, it is unclear what the rest of this reference means, or how it supports Lomborg's non-peer-reviewed claims about cold- and heat-related deaths due to global warming. Furthermore, the URL that Lomborg provided to locate the first reference (CRU, 2006) is an unidentifiable document.[19] And the second source he cites—"WHO, WMO, & UNEP, 2003:7"—does indeed report that "world temperature has increased by around 0.4°C since the 1970s," though the entire sentence he cited reads: "World temperature has increased by around 0.4°C since the 1970s, *and now exceeds the upper limit of natural (historical) variability*" (emphasis added)—undermining his claim that global warming is mostly the result of natural climate variability.[20] Furthermore, the same page of the same WHO report states:

> Change in world climate would influence the functioning of many ecosystems and their member species. Likewise, there would be impacts on human health. Some of these health impacts would be beneficial. For example, milder winters would reduce the seasonal winter-time peak in deaths that occurs in temperate countries, while in currently hot regions a further increase in temperatures might reduce the viability of disease-transmitting mosquito populations. *Overall, however, scientists consider that most of the health impacts of climate change would be adverse.*[21] [Emphasis added.]

In citing this page in the WHO report, Lomborg did not mention its conclusion that "overall . . . most of the health impacts of climate change would be adverse." Rather than mentioning this authoritative WHO assessment of the overall impact of global warming on human health, which appears on the same page of a document that Lomborg himself referenced, Lomborg argued to the contrary—without evidence—that the net impact of global warming on human health will be beneficial.

For those who might wonder why Lomborg would begin a section in *Cool It* titled "Malaria in Vermont" with an analysis of how the WHO "left out" key data revealing the health benefits of global warming, it was to present a case study of how the WHO's statistic of 150,000 deaths in 2000 because of global warming was supposedly exaggerated, just as the potential spread of malaria as a result of global warming is exaggerated.[22] Here again Lomborg identifies alleged culprits who, in this case, peddle the malaria "scare." Thus, Lomborg wrote: "Former UN secretary-general Kofi Annan told us 'a warmer world is one in which infectious diseases such as malaria and yellow fever will spread further and faster.'"[23] Although this quote is accurate, it should be put in the context of the speech Annan was delivering at the time to a climate conference in Nairobi, Kenya, in November 2006: "Climate change is not just an environmental issue, as too many people still believe. It is an all-encompassing threat. It is a threat to health, since a warmer world is one in which infectious diseases such as malaria and yellow fever will spread further and faster. It could imperil the world's food supply, as rising temperatures and prolonged drought render fertile areas unfit for grazing or crops. . . . This is not science fiction. These are plausible scenarios, based on clear and rigorous scientific modelling."[24] This is another example of the way Lomborg indicts respected groups and individuals—in this case the WHO, which he accused of spiking data, and the secretary-general of the United Nations, who, according to Lomborg, embellished facts—and thus makes them complicit in the supposedly broad phenomenon of exaggerated threats about the impacts of global warming.

A third victim in this case was a *U.S. News and World Report* cover story on global warming. Lomborg wrote, "The cover story from *U.S. News & World Report* that predicted waterlogged Art Deco hotels in Miami Beach also expected that in the future 'malaria could be a public health threat in Vermont.'"[25] The publication date of this article is January 28, 2001, which, as the piece itself noted, was shortly after the IPCC issued its 2001 assessment report on climate change. In its 2001 assessment the IPCC reported: "Vector-borne diseases, including malaria and dengue fever, may expand their ranges in the United States and may develop in Canada. Tick-borne Lyme disease also may see its range expanded in Canada. However, socioeconomic factors such as public health measures will play a large role in determining the existence or extent of such infections."[26] And: "Malaria was successfully eradicated from Australia, Europe, and the United States in the 1950s and 1960s, but the vectors were not eliminated. In regions where the vectors persist in sufficient abundance, there is a risk of locally transmitted malaria. This small risk of very localized outbreaks may increase under climate change."[27]

U.S. News also reported that "warmer weather would affect transmission of insect-borne diseases such as malaria and West Nile virus, which made a surprise arrival in the United States in 1999."[28] Based on what the IPCC reported in early January 2001, *U.S. News* did not appear to exaggerate when it reported in late January 2001 that malaria could become a public health threat in Vermont, because Vermont borders Canada, Vermont is in the United States, and the 2001 IPCC assessment noted that malaria could spread to the United States and Canada. Just as it mentioned the Art Deco hotels on Miami Beach, *U.S. News* simply invoked Vermont as a symbolic venue that clearly lies within the geographic range of a projected impact of global warming, as the IPCC reported.

For his part, Lomborg appeared to understate the potential spread of malaria as projected by his own sources (Martens et al., 1999; van Lieshout et al., 2004), and by the 2007 IPCC assessment. Shortly after charging *U.S. News* with exaggerating the threat of malaria, Lomborg wrote that the global "upper limit" of people at risk of malaria by year 2080 would be about 300 million.[29] But the Martens study that Lomborg cites did not project a single estimated upper limit; rather, it projected a range of additional people at risk of malaria from the mosquito-carried parasites (*P. falciparum* and *P. vivax*) from two climate models issued by the Hadley Center for Climate Prediction and Research in England. The Martens study reported:

"On a global level, the numbers of additional people at risk of malaria in 2080 due to climate change is estimated to be 300 and 150 million for *P. falciparum* and *P. vivax* types of malaria, respectively, under the HadCM3 climate change scenario. Under the HadCM2 ensemble projections, estimates of additional people at risk in 2080 range from 260 to 320 million for *P. falciparum* and from 100 to 200 million for *P. vivax*."[30] This means, according to Martens, that the range of the number of people at risk of malaria from climate change by 2080 would be from 360 million to more than 500 million, not an "upper limit" of 300 million, as Lomborg reported.

Also, van Lieshout, the second source cited by Lomborg, projected that the additional population at risk of malaria by 2080 would range from 220 million to more than 400 million, depending upon which SRES development scenario the study used in its estimates.[31] Thus, for van Lieshout, 400 million—not Lomborg's figure of 300 million—would be the upper limit. Likewise, the 2007 IPCC assessment, which Lomborg did not cite, reported: "Mixed projections for malaria are foreseen: globally an estimated additional population at risk between 220 million and 400 million has been estimated."[32] So for each of these three sources—two that Lomborg cited and one (the IPCC) that Lomborg should have cited—the "upper limit" of the number of people at risk of malaria by 2080 exceeded Lomborg's upper limit of 300 million by 100 to 200 million.

Furthermore, Lomborg's reference for Martens and van Lieshout reads in its entirety (parentheses in original): "(Martens et al., 1999; van Lieshout et al., 2004), behind the claims of (King, 2004)."[33] It is not clear why Lomborg cited the Martens and van Lieshout projections of susceptible malaria populations as being "behind the claims of King, 2004," especially considering the guest column that British scientist David A. King wrote for *Science*, and to which Lomborg refers. King, who then was the chief science adviser to the British government of Tony Blair, did not cite or otherwise refer to any malaria model in his *Science* column, nor did he provide any specific projections of susceptible populations to malaria. King's only reference to malaria reads: "As a consequence of continued warming, millions more people around the world may in future be exposed to the risk of hunger, drought, flooding, and debilitating diseases such as malaria." Odd that Lomborg would reference King, as these assertions seem contrary to what Lomborg reported in *Cool It*. In the same *Science* column, King also wrote: "In my view, climate change is the most severe problem that we are facing today—more serious even than the threat of terrorism." This also is at odds with Lomborg's notion of global warming as being no catastrophe. What's more, King wrote, again contrary to Lomborg: "It's a myth that reducing carbon emissions necessarily makes us poorer."[34] Lomborg does not explain why he would reference a source to appear as though it supported his argument, when in fact King's comments are clearly at odds with Lomborg's malaria-specific and broader views about global warming.

The projected global spread of malaria in response to climate change seems difficult to assess, especially given the absence of any definitive determination of the risk in the 2007 IPCC assessment report. However, Lomborg's indifference to the potential global spread of the disease—with uncertain consequences—is consistent with his seeming indifference toward the permanent loss of glaciers in the Himalayas as a source of fresh water for hundreds of millions of people. Lomborg's remedies—improved water storage in the Himalayas and mosquito nets[35]—though important as present-day and surrogate remedies in a climate-change context, are not wholly satisfactory substitutes for limiting the impact of climate change by reducing greenhouse emissions. Indeed, Lomborg clearly indicates his preferences: "I hope we will all work to make the best solutions [to global warming] cool. I would like to see college drives for mosquito nets against malaria before drives for adopting Kyoto."[36]

eleven
ON MALNUTRITION

Lomborg argues in "More Heat Means More Starvation?"—a section in *Cool It*—that the impact of global warming on human malnutrition is exaggerated. He wrote: "*Many people* worry that climate change will dramatically undermine our future ability to feed ourselves. *Stories* of how global warming will 'greatly increase the number of hungry people' and of how we are facing 'catastrophe' with 'whole regions becoming unsuitable for producing food' abound. Yes, global warming might slightly slow food production, but the claims are *vastly overplayed* and again—if our concern truly lies with food security and the world's hungry—lead us to focus on the wrong solutions" (emphasis added).[1]

Who are these excessively worried "people," and what are their "stories" that "vastly overplay" the idea that global warming could increase the number of malnourished people in the world? According to Lomborg's sources, the "people" are the British news service Reuters, and the British newspaper the *Independent*. And the "stories" are what Reuters and the *Independent* reported. Reuters summarized a report by the UN's Food and Agriculture Organization (FAO), and the *Independent* reported on a paper by Bill Hare of the Potsdam Institute for Climate Impact Research, which is Germany's leading global warming research center. Hare's paper, presented at the Hadley Center for Climate Prediction and Research in Exeter, England, as the *Independent* noted, was "a synthesis of a wide range of recent academic studies" that "pulls together for the first time the projected impacts on ecosystems and wildlife, food production, water resources and economies across the earth"[2] pursuant to projected increases in global temperatures in the twenty-first century. The *Independent* summarized Hare's February 2005 presentation:

> As present world temperatures are already 0.7°C above the pre-industrial level, the process is well under way. In the near future—the next 25

years—as the temperature climbs to the 1°C mark, some specialized ecosystems will start to feel stress, such as the tropical highland forests of Queensland, which contain a large number of Australia's endemic plant species, and the succulent karoo plant region of South Africa. In some developing countries, food production will start to decline, water shortage problems will worsen and there will be net losses in GDP.

It is when the temperature moves up to 2°C above the pre-industrial level, expected in the middle of this century—within the lifetime of many people alive today—that serious effects start to come thick and fast, studies suggest. Substantial losses of Arctic sea ice will threaten species such as polar bears and walruses, while in tropical regions "bleaching" of coral reefs will become more frequent—when the animals that live in the coral are forced out by high temperatures and the reef may die. Mediterranean regions will be hit by more forest fires and insect pests, while in regions of the U.S. such as the Rockies, rivers may become too warm for trout and salmon.

In South Africa, the Fynbos, the world's most remarkable floral kingdom which has more than 8,000 endemic wild flowers, will start to lose its species, as will alpine areas from Europe to Australia; the broadleaved forests of China will start to die. The numbers at risk from hunger will increase and another billion and a half people will face water shortages, and GDP losses in some developing countries will become significant.

But when the temperature moves up to the 3°C level, expected in the early part of the second half of the century, these effects will become critical. There is likely to be irreversible damage to the Amazon rainforest, leading to its collapse, and the complete destruction of coral reefs is likely to be widespread.

The alpine flora of Europe, Australia and New Zealand will probably disappear completely, with increasing numbers of extinctions of other plant species. There will be severe losses of China's broadleaved forests, and in South Africa the flora of the Succulent Karoo will be destroyed, and the flora of the Fynbos will be hugely damaged.

There will be a rapid increase in populations exposed to hunger, with up to 5.5 billion people living in regions with large losses in crop production, while another 3 billion people will have increased risk of water shortages.

Above the 3°C raised level, which may be after 2070, the effects will be catastrophic: the Arctic sea ice will disappear, and species such as

> polar bears and walruses may disappear with it, while the main prey species of Arctic carnivores, such as wolves, Arctic foxes and the collared lemming, will have gone from 80 per cent of their range, critically endangering predators.
>
> In human terms there is likely to be catastrophe too, with water stress becoming even worse, and whole regions becoming unsuitable for producing food, while there will be substantial impacts on global GDP.[3]

Lomborg's representation of Hare's paper was to call it a "story" and to comment (without engaging Hare's summary): "Yes, global warming might slightly slow food production, but the claims are vastly overplayed."[4]

Lomborg likewise responded to the Reuters piece, published in May 2005, which summarized the FAO report (and which appeared to corroborate Hare's analysis three months earlier): "Global warming is likely to significantly diminish food production in many countries and greatly increase the number of hungry people, the U.N. Food and Agriculture Organization said Thursday. FAO said in a report food distribution systems and their infrastructure would be disrupted and that the severest impact would likely be in sub-Saharan African countries."[5] The report continued: "FAO said scientific studies showed that global warming would lead to an 11 percent decrease in rainfed land in developing countries and in turn a serious decline in cereal production. Sixty-five developing countries, representing more than half of the developing world's total population in 1995, will lose about 280 million tons of potential cereal production as a result of climate change."

The effect of climate change on agriculture could increase the number of people at risk of hunger, particularly in countries already saddled with low economic growth and high malnourishment levels. "In some 40 poor, developing countries, with a combined population of 2 billion . . . production losses due to climate change may drastically increase the number of undernourished people, severely hindering progress in combating poverty and food insecurity," the report said.[6] Lomborg ignored all of this seemingly relevant information published in Reuters and the *Independent*. Instead, while citing these two reports, he wrote, "To put the issue in context, food availability has increased dramatically over the past four decades."[7] However, neither of his referenced sources focused on "the past four decades"; rather, they were focused on what would happen in the future when the average global temperature increases by 1, 2, and 3°C as the result of global warming (the *Independent*'s coverage of Hare), and the future impact on food and water security in a warming world (Reuters's coverage of FAO).

In much the same way that he presented his own ad hoc projections of sea-level rise this century,[8] Lomborg furnishes his own world-hunger report separate from the existing peer-reviewed scientific research. Lomborg wrote: "The average person in the developing world has experienced a 40 percent increase in available calories."[9] Even if this undocumented assertion were accurate, it has little relevance to global warming projections and future access to food and water. Lomborg continued: "Likewise, the proportion of malnourished has dropped from 50 percent to less than 17 percent."[10] Here again, his statement is irrelevant to the twenty-first century in the context of global warming. Although this sentence cited seven sources, none seem to support this specific assertion.[11] He went on: "The UN expects these positive trends to continue at least till 2050 with another 20 percentage points' calorie increase and malnourished dropping below 3 percent"[12] (no sources were included for this sentence).

While providing this backward-looking assessment of the future impact of global warming on the availability of food and water, Lomborg ignored the 2007 IPCC assessment, which provided nuanced projections of the impact of global warming on food production and malnutrition. With respect to food production, the IPCC reported:

> [Climate] modelling results for a range of sites find that, in mid-to high-latitude regions, moderate to medium local increases in temperature (1–3°C), along with associated carbon dioxide (CO_2) increase and rainfall changes, can have small beneficial impacts on crop yields. In low-latitude regions, even moderate temperature increases (1–2°C) are likely to have negative yield impacts for major cereals. Further warming has increasingly negative impacts in all regions (medium to low

> confidence). These results, on the whole, project the potential for global food production to increase with increases in local average temperature over a range of 1 to 3°C, but above this range to decrease.[13]

The 2007 IPCC assessment also reported:

> Globally, the potential for food production is projected to increase with increases in local average temperature over a range of 1 to 3°C, but above this it is projected to decrease. Changes in the patterns of extreme events, such as increased frequency and intensity of droughts and flooding, will affect the *stability* of, as well as *access* to, food supplies. Food insecurity and loss of livelihood would be further exacerbated by the loss of cultivated land and nursery areas for fisheries through inundation and coastal erosion in low-lying areas.[14] [Emphasis in original.]

With respect to malnutrition, the 2007 IPCC assessment report concluded: "The projected relative risks attributable to climate change in 2030 show an increase in malnutrition in some Asian countries. Later in the century, expected trends in warming are projected to decrease the availability of crop yields in seasonally dry and tropical regions. This will increase hunger, malnutrition and consequent disorders, including child growth and development, in particular in those regions that are already most vulnerable to food insecurity, notably Africa."[15] And finally, also with respect to malnutrition: "Agricultural production in many African countries and regions will likely be severely compromised by climate change and climate variability. This would adversely affect food security and exacerbate malnutrition (very high confidence)."[16]

Lomborg goes on to discuss various surveys pertaining to climate change and food production: "A few large-scale surveys have looked at the effect of climate change on agricultural production together with the global food-trade system. There are four crucial findings generally shared among them."[17] He referenced six such surveys,[18] and of these four "crucial" findings he first notes that "all models envision a large increase in agricultural output—more than a doubling of cereal production over the coming century. Thus, we will be able to feed the world ever better."[19] Although the models in these surveys envision an increase in agricultural production in the wealthy developed world, the prognosis is very different for the poorer developing world. One of the six surveys (Parry et al., 2004), while referring to the development scenario that would lead to the highest temperature increase because of global warming, concluded: "The A_1F_1 scenario, as expected with its large increase in global temperatures, exhibits the greatest decreases both regionally and globally in yields, especially by the 2080s."[20]

The Parry paper also noted a severe asymmetry between agricultural production in the developed world as opposed to the developing world—with a net global increase in agricultural production, as Lomborg observed, but with "regional differences in crop production [that] are likely to grow stronger through time, leading to a significant polarisation of effects, with substantial increases in prices and risk of hunger amongst the poorer nations, especially under scenarios of greater inequality."[21]

Lomborg's five other sources generally concurred with this projection of globally higher food production in the context of severe regional asymmetries. The study of climate models that Lomborg identified as his principal source for this section (Fischer et al., 2005)[22] reported that "critical impact asymmetries . . . may deepen current production and consumption gaps between developed and developing world."[23] The same paper also reported: "Some fairly robust conclusions emerge from the analysis of climate-change impacts on the number of people at risk of hunger. *First, climate change will most likely increase the number of people at risk of hunger*. Second, the importance and significance of the climate-change impact on the level of undernourishment depends entirely on the level of economic development assumed in the SRES scenarios"[24] (emphasis added). Though Lomborg noted that the Fischer paper was "the central one used here" (among the six sources that he referenced) to support the "good news" about global warming and human malnutrition,[25] he neglected to mention that the same paper reported the "fairly robust conclusion" that "climate change will most likely increase the number of people at risk of hunger."

Another of Lomborg's sources (Fischer et al., 2002) also noted that the developing world will disproportionately suffer the consequences of global warming when it comes to food production. Summarizing the results of a set of climate models (HadCM3), this document reported:

> For four HadCM3 climate-change scenarios, the estimated impacts on rain-fed cereal-production potential on current cultivated land imply that there are 42–73 countries with potential cereal-productivity declines of more than 5% ("losing" countries). The population in 2080 of these countries ranges between 1.6 billion and 3.8 billion people. In these countries, cereal-production losses amount to 3–8% of the global potential; a grim outlook for the already poor among these losing countries despite substantial increases in some 54–71 gaining countries. . . .
>
> Individual country results are reason for concern. For example, in the results of the HadCM3 scenarios, 20–40 poor and food-insecure countries, with a projected total population in 2080 in the range of 1–3 billion, may lose on average 10–20% of their cereal-production potential due to climate change.[26]

Another of Lomborg's sources (Fischer, van Velthuizen, Shah, and Nachtergaele, 2002), after reviewing several climate models, also reported projected disparities between developed and developing countries: "The results highlight that climate change will benefit the developed countries more than the developing countries regardless of what is assumed, when considering one rain-fed cereal crop per year or for multiple rain-fed cropping and irrigated production. Also, the results clearly demonstrate that climate impacts will be heterogeneous and vastly different across regions, with the potential of putting major burdens on some 40 to 60 developing countries."[27] Another source (Parry, Rosenzweig, and Livermore, 2005) reported, like the others: "Generally, the SRES scenarios result in crop yield decreases in developing countries and yield increases in developed countries. . . . Decreases are especially significant in Africa and parts of Asia with expected losses up to 30%."[28] And the sixth and last cited source (Rosenzweig and Parry, 1994) reported that "developing countries are likely to bear the brunt of the problem, and simulations of the effect of adaptive measures by farmers imply that these will do little to reduce the disparity between developed and developing countries."[29]

These sources—which Lomborg cited but did not review in any detail—do not support his first finding that "we will be able to feed the world ever better" as a result of global warming.[30]

Lomborg proceeded to report on the other three findings:

> Second, the impact of global warming on food production will probably be negative but in total very modest. . . . Third, while globally there will be very little change, regionally this is not true. Global warming in general has a negative impact on third-world agriculture, whereas it has a positive impact on first-world farming. This is because temperature increases are good for farmers in high latitudes, where more warmth will lead to longer growing seasons, multiple harvests, and higher yields. For farmers in tropical countries—typically, third-world countries—higher temperatures mean lower agricultural productivity. For both places, however, CO_2 in itself counts as a positive factor, since it acts as a fertilizer, making crops grow more everywhere. . . . Fourth, global warming will mean slightly more malnourished people, because food production will decline slightly.[31]

Lomborg does not reconcile the opposing conclusions between his first finding—the "good news" that "we will be able to feed the world ever better"—and his second, third, and fourth, including that the world will have "slightly more malnourished people."

In assessing whether global warming would enable the world to be better fed, Lomborg neglected to cite the 2007 IPCC assessment:

> Projected trends in climate-change related exposures of importance to human health will have important consequences (high confidence). Projected climate-change related exposures are likely to affect the health status of millions of people, particularly those with low adaptive capacity, through: increases in malnutrition and consequent disorders, with implications for child growth and development . . .[32]
>
> Agricultural production in many African countries and regions will likely be severely compromised by climate change and climate variability. This would adversely affect food security and exacerbate malnutrition (very high confidence).[33]

> Climate change poses substantial risks to human health in Asia. Global burden (mortality and morbidity) of climate-change attributable diarrhoea and malnutrition are already the largest in South-East Asian countries including Bangladesh, Bhutan, India, Maldives, Myanmar and Nepal in 2000, and the relative risks for these conditions for 2030 is expected to be also the largest, although in some areas, such as southern states in India, there will be a reduction in the transmission season by 2080.[34]

At best, Lomborg's analysis demonstrated that the scale and complexity of the issue of the future impact of global warming on human malnutrition is beyond the capacity of one individual to credibly present in a few hundred words as a non-peer-reviewed alternative to the comprehensive, peer-reviewed 2007 IPCC assessment report.[35] It is also fair to say that Lomborg's prognosis that "we will be able to feed the world ever better" because of global warming was not supported by either Lomborg's own sources or the 2007 IPCC assessment of the issue.

twelve
ON WATER SHORTAGES

Lomborg argued that global warming will reduce the number of people worldwide who lack access to drinking water. In this context, Lomborg indicted "environmental circles, where the argument is that we're approaching a [water] crisis" and a "full-scale emergency," which, according to Lomborg, is "misleading." Lomborg acknowledged "regional and logistic problems with water," and that "we need to get better at using it." "But basically," he writes, "we have sufficient water."[1] His conclusions are inconsistent with the 2007 IPCC assessment: "Currently, human beings and natural ecosystems in many river basins suffer from a lack of water. . . . These basins are located in Africa, the Mediterranean region, the Near East, South Asia, Northern China, Australia, the USA, Mexico, north-eastern Brazil, and the western coast of South America. Estimates of the population living in such severely stressed basins range from 1.4 billion to 2.1 billion."[2] And: "Australia, western USA and southern Canada, and the Sahel have suffered from more intense and multi-annual droughts, highlighting the vulnerability of these regions to the increased drought occurrence that is expected in the future due to climate change."[3]

Similarly, while Lomborg accurately quoted from a 2006 United Nations Educational, Scientific, and Cultural Organization (UNESCO) paper,[4] which reported that lack of water "is primarily driven by an inefficient supply of services rather than by water shortages,"[5] the same UNESCO document reported: "In 2000, of the world's total population 20% had no appreciable natural water supply, 65% shared low-to-moderate supplies, and only 15% enjoyed relative abundance."[6] These assessments would seem to indicate that we do not have sufficient water, and that global warming will exacerbate widespread water insufficiency.

In the endnote supporting his statement that lack of water "is primarily driven by an inefficient supply of services rather than by water shortages," Lomborg added, "Also, 'there is enough water for everyone. The problem we face today is largely one of governance' (3)."[7] The "3" in parentheses presumably indicates that the words in quotation marks are referenced to page 3 of the same UNESCO document referenced in the endnote (which is the only reference cited); however, no such words appear on page 3 or anywhere else in the UNESCO document.[8]

Lomborg also gives a specific cost estimate of providing drinking water and sanitation to those in the world who lack these essential services (one billion lack access to drinking water, and two and a half billion lack access to sanitation). He wrote: "We could bring basic water and sanitation to all of these people within a decade for about $4 billion annually."[9] However, in the endnote supporting the $4 billion per year figure, Lomborg wrote: "About $10 billion per year from 2007 to 2015, from a range of global studies (Toubkiss, 2006:7)."[10] Without citing any additional sources, Lomborg wrote in the same endnote, "Four billion dollars per year forever is the equivalent to $10 billion from 2007 to 2015 at a 5 percent discount rate."[11] Lomborg clearly would have done better to give the present-value estimate of $10 billion, if that in fact is what his sources had reported. In fact, the source that Lomborg referenced to support his claim that "we could bring basic water and sanitation to all of these people within a decade for about $4 billion annually," did not give an annual estimate of $4 billion. The source he referenced here, the World Water Council report "Costing MDG Target 10 on Water Supply and Sanitation (Toubkiss, 2006),"[12] provided a range of estimates—*beginning* at $10 billion per year:

> The range of the estimates
>
> At first, the range of the global estimates seems broad—between 9 billion USD (WHO 2004) and 30 billion USD per year (GWP 2000 and World Bank 2003). After closer examination however, a different picture emerges. Indeed, if the results are analysed on comparable bases, they appear quite similar: approximately 10 billion USD per year would be required to supply low-cost water and sanitation services to people who are not currently supplied (WSSCC 2000, WHO 2004), a

> further 15 to 20 billion USD a year to provide them with a higher level of service and to maintain current levels of service to people who are already supplied (Water Academy 2004). A much larger figure, up to 80 billion USD, is projected solely for collecting and treating household wastewater and for preserving the global environment through integrated water resources management (IWRM) and ecological methods (GWP 2000 and SEI 2004).[13]

Lomborg's estimate of $4 billion per year appears neither in the "Target 10" report he cited nor within the range of estimates (between "10 billion USD" and "80 billion USD" per year) that the Target 10 report considered.

Lomborg argued that "the future water challenge lies not primarily in regulating global warming"—that is, in reducing greenhouse emissions—"but in ensuring that three billion people can get access to clean drinking water and sanitation."[14] This statement incorrectly implies that advocates of reducing greenhouse emissions believe that reducing these emissions would be sufficient to address current and future water challenges. As he did throughout *Cool It*, Lomborg then closed this section by repeating an unsupported assertion as though it were fully documented, in this case that it would cost only $4 billion annually through year 2015 to supply fresh water and sanitation to everyone in the world who needed it: "The future water challenge lies not primarily in regulating global warming but in ensuring that three billion people can get access to clean drinking water and sanitation. This small policy change would be remarkably inexpensive at $4 billion annually and would bring huge health and quality-of-life benefits to half the world's population."[15]

So ended the presentation of "Lomborg's Theorem" in *Cool It*, wherein Lomborg argued that global warming is "no catastrophe" and consequently requires no serious global effort to reduce greenhouse emissions.

PART 3

Journalism as Usual

thirteen
LOMBORG'S TRIPLE-A RATING

The Intergovernmental Panel on Climate Change was formed by the United Nations Environment Program and the UN's World Meteorological Organization in 1988. According to its statement of principles, the role of the IPCC is "to assess on a comprehensive, objective, open and transparent basis the scientific, technical and socio-economic information relevant to understanding the scientific basis of risk of human-induced climate change, its potential impacts and options for adaptation and mitigation." IPCC reports "should be neutral with respect to policy, although they may need to deal objectively with scientific, technical and socio-economic factors relevant to the application of particular policies." The IPCC "also shall reflect balanced geographic representation" and "in taking decisions, and approving, adopting and accepting reports . . . shall use all best endeavours to reach consensus."[1]

Given the emphasis in this statement on scientific objectivity, policy neutrality, balanced geographic representation, and consensus, one might conclude that by the time its four major assessment reports were issued—in 1990, 1995, 2001, and 2007—the IPCC's peer-reviewed product would embody a scientifically sound consensus middle ground among its 2,500 contributors and reviewers. Consequently, it is difficult to fathom how Lomborg's *Cool It*, which reflects none of these characteristics and which throughout asserts unsubstantiated claims that are completely at odds with the published consensus among the IPCC-affiliated scientists, would be described by the leading conservative newspaper in the United States as representing "the practical middle" (*Wall Street Journal*) and by the leading liberal newspaper as representing "the pragmatic center" (*New York Times*) on global warming.[2] Although Lomborg is responsible for what he wrote, he is not wholly responsible for his success, much of which can be attributed to the poorly researched reviews of his books by the leading newspapers and journals in the United States.

Kimberly Strassel wrote among the first of these reviews for the *Wall Street Journal* in September 2007. Strassel, a member of the *Journal*'s editorial board, seemed content to take Lomborg's factual assertions at face value without apparently consulting the 2007 IPCC assessment report, which was issued a few months earlier. Instead, Strassel was struck by "the free-thinking Dane" and his new book, which is "brimming with useful facts and common sense." She wrote: "Mr. Lomborg starts by doing what he does best: presenting a calm analysis of what today's best science tells us about global warming and its risks. Relying primarily on official statistics, he ticks through the many supposed calamities that will result from a hotter planet—extreme hurricanes, flooding rivers, malaria, heat deaths, starvation, water shortages. It turns out that, when these problems are looked at from all sides and stripped of the spin, they aren't as worrisome as global-warming alarmists would suggest. In some cases, they even have an upside."[3] The three major elements of this review—Lomborg's authoritative "facts," his "centrist" position, and his "calm" tone—would all be featured in Andrew Revkin's review of *Cool It* in the *New York Times* soon afterward.

A month before Revkin's review appeared, however, Lomborg was allowed to write, in effect, his own review of his newly published book in the *Washington Post*. Like Strassel and, later, Revkin, Lomborg argued that his book staked out "the middle ground, where we can have a sensible discussion." Lomborg presented his book as inarguably factual, while urging readers to "Chill Out" (the title of his piece in the *Post*) about the threat of global warming. By acknowledging the reality of global warming or the fact that it's caused in large part by humans, he could proceed to position himself as a "centrist" by asserting that global warming is no catastrophe: "The discussion about climate change has turned into a nasty dustup, with one side arguing that we're headed for catastrophe and the other maintaining that it's all a hoax. I say that neither is right. It's wrong to deny the obvious: The Earth is warming, and we're causing it. But that's not the whole story, and predictions of impending disaster just don't stack up."[4]

Just as Lomborg did, the *Times*'s global warming reporter, Andrew

Revkin, began his review by simplistically framing the debate in the United States about global warming as "a yelling match between the political and environmental left and the right." Similar to Strassel's and Lomborg's articles, Revkin situated *Cool It* in the "pragmatic center" and in the "centrist camp" among other books on global warming. And Revkin too signed on to Lomborg's authoritativeness while erring in his only mention of a substantive issue pertaining to *Cool It:* "His first book, 'The Skeptical Environmentalist,' put him on Time magazine's list of 100 most influential people in 2004 and made him a star among conservative politicians and editorial boards. In his short new book, 'Cool It: The Skeptical Environmentalist's Guide to Global Warming,' Mr. Lomborg reprises his earlier argument with a tighter focus. He tries to puncture more of what he says are environmental myths, like the imminent demise of polar bears." Revkin then claimed: "Most bear biologists have never said the species is doomed but do see populations shrinking significantly in a melting Arctic."[5]

In his unquestioning acceptance of Lomborg's assessment, Revkin neglected to acknowledge the 2004 Arctic Climate Impact Assessment (ACIA), which reported: "It is difficult to envisage the survival of polar bears as a species given a zero summer sea-ice scenario."[6] Approximately three hundred scientists contributed to the 2004 ACIA report, so it is difficult to reconcile Revkin's claim regarding the views of "most bear biologists." In addition, and separate from the report, leading polar bear scientist Ian Stirling clearly links the impact of global warming on shrinking Arctic sea ice as determinative of the polar bears' future: "Polar bears evolved into existence because of a large productive habitat (sea ice), unoccupied by a terrestrial predator. As that habitat disappears, so eventually will the bears that live and depend on it."[7]

Revkin's *New York Times* colleague John Tierney, who writes frequently about global warming, also wrote about *Cool It.* In September 2007, on the occasion of the book's publication, Tierney visited the Bridge Café near the Brooklyn Bridge in New York "accompanied by Bjorn Lomborg, the Danish political scientist and scourge of environmentalist orthodoxy." Tierney sardonically observed that the café was unaffected to date by sea-level rise from global warming: "We couldn't see any evidence of the higher sea level near the Bridge Café, mainly because Water Street isn't next to the water anymore. Dr. Lomborg and I had to walk over two-and-a-half blocks of landfill to reach the current shoreline." Tierney likewise contended that global warming is no catastrophe and agreed with Lomborg that global warming would lead to a net reduction in human mortality because there would be fewer cold-related deaths. And, like Lomborg, Tierney did this while throwing a quick jab at Al Gore:

> Hotter summer can indeed be fatal, as Al Gore likes us to remind audiences by citing the 35,000 deaths attributed to the 2003 heat wave in Europe. But there are a couple of confounding factors explained in Dr. Lomborg's new book, "Cool It: The Skeptical Environmentalist's Guide to Global Warming."
>
> The first is that winter can be deadlier than summer. About seven times more deaths in Europe are attributed annually to cold weather (which aggravates circulatory and respiratory illness) than to hot weather, Dr. Lomborg notes, pointing to studies showing that a warmer planet would mean fewer temperature-related deaths in Europe and worldwide.[8]

Like Revkin, Tierney unquestioningly accepted Lomborg's argument. If he had checked, he might well have come across the "Human Health" chapter in the 2007 IPCC assessment report, which extensively examined the direct and indirect impacts of rising temperatures on human mortality.[9] This part of the report concluded that global warming will "bring some benefits to health, including fewer deaths from cold, although it is expected that these will be outweighed by the negative effects of rising temperatures worldwide, especially in developing countries (high confidence)."[10] Instead Tierney, like Strassel and Revkin, chose to promote Lomborg's "practical" (Strassel), "pragmatic" (Revkin), "middle" (Strassel), and "centrist" (Revkin) analysis.

Lomborg was also favorably reviewed in the leading U.S. foreign-policy journal, *Foreign Affairs.* In his brief review Richard N. Cooper, a chaired professor of international economics at Harvard University and a member of the *Foreign Affairs* book review panel,[11] wrote: "The title of this highly readable book has a double meaning: steps should be taken against global warming, but unsupported claims that climate change will lead to global catastrophe and human calamity should be avoided." Cooper, like the three journalists, simply conveyed Lomborg's analysis as if it were a well-supported dissent from the scientific consensus. For example, Cooper wrote: "Lomborg also debunks some of the more spectacular claims about climate change—for example, that it is depleting the global population of polar bears."[12] Even putting aside that the polar bear population is not often assessed on a "global" basis, given that they live only in the Arctic, it is disconcerting that none of these high-end commentators ever bothered to point out the contradictions between Lomborg's writings and the 2004 Arctic Climate Impact Assessment, the reports by the IUCN's Polar Bear Specialist Group, and the IPCC assessment reports.

The newspapers and journals that reviewed *Cool It* while not bothering to investigate Lomborg's science-related assertions also neglected to mention that Lomborg's opposition to a significant reduction in greenhouse emissions lies outside the global political consensus, as stipulated in the 1992 United Nations Framework Convention on Climate Change (UNFCCC)—which the United States signed and ratified, and which states:

> The ultimate objective of this Convention and any related legal instruments that the Conference of the Parties may adopt is to achieve, in accordance with the relevant provisions of the Convention, stabilisation of greenhouse gas concentrations in the atmosphere at a level that would prevent dangerous anthropogenic interference with the climate system. Such a level should be achieved within a time frame sufficient to allow ecosystems to adapt naturally to climate change, to ensure that food production is not threatened and to enable economic development to proceed in a sustainable manner.[13]

The U.S. Constitution (Article VI, Section 2) makes U.S.-ratified treaties the "supreme law of the land" in the United States, and the

1992 UNFCCC is a U.S.-ratified treaty. So given that the main stipulation of that treaty requires the United States as a state signatory to contribute to the stabilization of greenhouse emissions to an extent necessary to "prevent dangerous anthropogenic interference with the climate system," and given that the United States historically is the world's most prolific generator of greenhouse emissions, it would not have been unreasonable to expect at least one leading U.S. newspaper or journal to point out that Lomborg's supposedly admirable "centrism" in opposing significant greenhouse emissions, in addition to lying outside the scientific and environmental consensus, also exists outside a consensus political and legal mandate to reduce greenhouse emissions.

As recently as February 2009, the *Times*'s Tierney sought to promote Lomborg's work as if it were scientifically credible, despite the emergence of evidence after the September 2007 publication of *Cool It* showing the potential impacts of global warming to be even more serious than conveyed in the 2007 IPCC assessment report. This discussion appeared as part of an item Tierney published on February 23, 2009, for his "Tierney Lab" blog for the *New York Times*, questioning the scientific integrity of Harvard professor John Holdren, who had been nominated by President Barack Obama to be the president's chief science adviser.[14] Soon after Holdren's Senate confirmation hearings on February 12, but before the confirmation vote by the Senate, Tierney cited a 2007 book by Roger A. Pielke Jr., in which Pielke accused Holdren—in addition to other prominent scientists, including Stanford University's Stephen Schneider, the Heinz Center's Thomas Lovejoy, and Duke University's Stuart Pimm—of politicizing science in their criticism of Lomborg's 2001 book, *The Skeptical Environmentalist*.[15] Tierney seemed to convey that Pielke's book cast doubt on Holdren's fitness to be the president's chief science adviser.[16]

In *The Honest Broker: Making Sense of Science in Policy and Politics* (published by Cambridge University Press, which was also publisher for *The Skeptical Environmentalist*), Pielke argued that the forums organized in late 2001 by the Union of Concerned Scientists and early 2002 by *Scientific American* to address major concerns pertaining to Lomborg's *The Skeptical Environmentalist* were primarily political exercises that reflected the scientists' dislike of Lomborg. Tierney wrote:

> What kind of advice should scientists give to politicians? I take up that question in my Findings column about Roger Pielke Jr.'s book, "The Honest Broker: Making Sense of Science in Policy and Politics," which argues that too many scientists stealthily pass off their own political views as incontestable scientific truths. As a result, they're tempted to try to win political debates by hyping their own expertise and denigrating their opponents as "unscientific."
>
> One of the "stealth issue advocates" discussed in the book is John P. Holdren, the Harvard physicist who is awaiting confirmation as Mr. Obama's science adviser. . . .
>
> But I share Dr. Pielke's concern about some of the debating tactics used by Dr. Holdren and his allies. Dr. Holdren began his career collaborating with the ecologist Paul Ehrlich, a master of the apocalyptic forecast and the contemptuous argument from authority. . . .
>
> Dr. Ehrlich and Dr. Holdren were was [*sic*] so confident in their expertise that they accepted Dr. [Julian] Simon's challenge to bet about the future price of natural resources. They wagered $1,000—and lost decisively. But that loss didn't diminish the neo-Malthusians' contempt for the other side when Dr. Simon's arguments were updated in 2001 by Bjorn Lomborg. As Dr. Pielke notes in his book, Dr. Holdren joined in an unusual effort by scientists to denounce Dr. Lomborg as unscientific. Dr. Holdren accused Dr. Lomborg of "complete incompetence," complained that Dr. Lomborg had "wasted immense amounts of the time of capable people," and labeled his ideas "dangerous for the future of society."
>
> What lessons do you draw from Dr. Pielke's book and from Dr. Holdren's past? And what kind of advice should Dr. Holdren and other scientists be giving to Mr. Obama?[17]

Pielke's book, like Tierney's blog entry, simply assumes that Lomborg's work is factually and methodologically sound. Unlike the scientists mentioned—Holdren, Schneider, Lovejoy, Pimm, and other major scientists—neither Pielke nor Tierney apparently sought to determine the factual and analytic integrity of Lomborg's book. And

Pielke presents little substantive analysis of the scientists' criticism of Lomborg. Despite these significant shortcomings in his own analysis, Pielke wrote in *The Honest Broker:* "In the case of *TSE* [*The Skeptical Environmentalist*], scientists served as Stealth Issues Advocates when they claimed that Lomborg had gotten his 'science' wrong, and because he has his science wrong then necessarily those who accept his views of 'science' should lose out in the political battle. . . . The debate [about Lomborg] was about political power and scientists readily chose sides as Issues Advocates."[18] For Pielke, then, the scientists who contributed to the *Scientific American* and UCS forums "served as Stealth Issues Advocates when they claimed that Lomborg had gotten his 'science' wrong." But could the scientists' credibly be charged as "Stealth Issues Advocates" if in fact Lomborg's science was radically wrong? And if Lomborg's science is wrong—or worse—about issues of monumental public concern, shouldn't Lomborg's scientist-critics be lauded for their citizenship, including, one might expect, in the most authoritative newspaper in the United States?

If the science increasingly shows—as the next chapter demonstrates that indeed it does—that global warming is a looming planetary disaster, what would Tierney and Pielke have these scientists do in response to Lomborg's plainly fraudulent theorem that global warming is no catastrophe? If anything, the Tierney-Pielke writings on Lomborg and his critics are symptomatic of how Lomborg's Theorem has infected the public discussion of global warming, even at the highest levels of journalism and book publishing.

Despite the twenty-year scientific assessment effort of the UN's Intergovernmental Panel on Climate Change, Lomborg successfully competed with the IPCC in the United States as the "skeptical environmentalist" who argued that global warming wouldn't be as bad as what the IPCC-projected impacts showed, and that greenhouse emissions should not be reduced to the extent that the UN's Framework Convention on Climate Change mandated. When Al Gore fully emerged in 2006 as the antidote to Lomborg on the basis of his film and book *An Inconvenient Truth*, Lomborg devoted much effort in

Cool It to discrediting Gore, with applause from news organizations and journalists in the United States.

The cultural and political narrative of an alleged Greenpeace-affiliated vegetarian to "Cool It" and "Chill Out" about global warming apparently was too compelling for the major U.S. news organizations to resist. Yet the favorable coverage of Lomborg and his books are to global warming what the triple-A ratings for mortgage-backed securities were to the U.S. financial system—misguided seals of approval with catastrophic consequences. Even worse, financial systems and economies presumably can be reinvented and restored, but the Earth, its climate, and its environment—upon which economic well-being and human civilization ultimately depend—cannot. Lomborg's success largely reflects an ability of elite publishing houses and news organizations to construct an alternative but counterfeit network of knowledge about an issue of the highest public importance.

fourteen
HOW WRONG WAS LOMBORG?

Not only were Lomborg's arguments poorly supported by the data available at the time, they also have not stood up to subsequent events. In May 2008, eight months after the publication of *Cool It*, in which he criticized Al Gore and *Time* for writing that polar bears may be endangered because of global warming, the Bush administration's Department of the Interior announced that it was listing polar bears as a threatened species under the U.S. Endangered Species Act "based on the best available science, which shows that loss of sea ice threatens and will likely continue to threaten polar bear habitat."[1] This assessment is very similar to the Artic Climate Impact Assessment report in 2004, which Lomborg disregarded. And in August 2008 — nearly a year after Lomborg mocked Gore for writing that "polar bears have been drowning in significant numbers" — the Center for Biological Diversity (CBD) in the United States issued a press release reporting that "recent government surveys document that polar bears are at risk of drowning in large numbers off the north coast of Alaska as sea ice once again approaches record low levels."[2] The center's 2008 press release continued: "In the rule listing the polar bear under the Endangered Species Act, the U.S. Fish and Wildlife Service concluded that polar bear drowning events are expected: 'As changes in habitat become more severe and seasonal rates of change more rapid, catastrophic mortality events that have yet to be realized on a large scale are expected to occur.' The highly unusual observation this week of so many bears in open water is cause for concern that such events are already occurring."[3]

These more recent statements, in addition to the 2001 analysis of the IUCN's Polar Bear Specialist Group and the 2004 Arctic Climate Impact Assessment discussed in chapter 1, leave little doubt that polar bear drownings and their ultimate survival is ineluctably linked to the condition of Arctic sea ice. Yet the words "Arctic sea ice" or "sea ice" make no appearance in Lomborg's chapter on polar bears, in which he argues that "we hear vastly exaggerated and emotional claims" about the threat to polar bears from global warming "that are simply not supported by data."[4]

And the situation regarding Arctic sea ice is deteriorating. Study after study that has been published since *Cool It* first appeared in September 2007 has demonstrated that this essential polar bear habitat is rapidly shrinking because of global warming. In December 2007, a report presented at a meeting of the American Geophysical Union concluded that "their latest modelling studies indicate polar waters could be ice-free in summers within just 5–6 years," and that summer melting in 2007 "reduced the ice cover to 4.13 million sq km, the smallest ever extent in modern times."[5] In January 2008 researchers reported in *Geophysical Research Letters*, "A new study using satellite measurements of Arctic sea ice has revealed that thinner ice that's only two or three years old now accounts for 58 percent of the ice cover—up from 35 percent in the mid-1980s. Meanwhile, ice older than nine years had all but disappeared by 2007. The extinction of the older, thicker ice is effectively melting away the Arctic Ocean's hedge against complete summer meltdowns, say researchers. 'The thinning is consistent with long-term warming,' said ice researcher James Maslanik of the University of Colorado in Boulder."[6]

In August 2008, German scientists reported in *Geophysical Research Letters* that "in 2007 the sea ice at the North Pole was at its thinnest since records began." The *New Scientist*, which assessed the study, reported that "while the ice at the North Pole used to be thick 'old' ice, much of it now is thinner first-year ice, which has had only a single winter to grow."[7] And in December 2008, according to the Australian Broadcasting Corporation, "scientists from Canada and more than a dozen other countries reported that the Arctic region will have an ice-free summer in as little as six years."[8]

In Fall 2007, two leading polar bear scientists, Ian Stirling and Andrew Derocher, made an explicit link between the survival of the polar bears and the shrinking Arctic sea ice. Speaking of the proposed listing of polar bears as a threatened species in January 2007 by the

U.S. Fish and Wildlife Service, Stirling and Derocher wrote that "habitat loss of sea ice is the central justification for the proposed listing." They continued:

> Contrarian articles continue to appear in the popular press, questioning climate warming in general and, more specifically, denying the potential negative effects on polar bears. Such articles generally exhibit a poor understanding of polar bear ecology and selectively use information out of context, which results in public confusion about the real threat to polar bears due to loss of sea ice. . . . In the long term, the loss of an iconic species such as the polar bear is but a symbol of much larger and hugely significant changes that will occur in many ecosystems throughout the world if the climate continues to warm and especially if, as projected by the IPCC, such warming is largely a consequence of excess anthropogenic [man-made] productivity of greenhouse gases. For polar bears, habitat loss is the most critical single concern. The symptoms of climate warming on polar bears are becoming clearer. Highly specialized species are particularly vulnerable to extinction if their environment changes, and polar bears fit that prescription. If the population of the planet is truly concerned about the fate of this species, we need to collectively reduce greenhouse gas production significantly and quickly.[9]

Clearly the scientific community is in agreement that shrinking Arctic sea ice as a result of global warming is "the most critical single concern" regarding the fate of the polar bear. It is thus quite incredible that Lomborg, and his supporters in the news media, would charge environmentalists with "vastly exaggerating" the threat to polar bears without an analysis of the shrinking polar bear sea-ice habitat.

Many other studies published after *Cool It* confirm that Lomborg was wrong on virtually every major claim that he made about supposed exaggerated threats of global warming. In *Cool It* Lomborg argued that the issue of melting glaciers, among other environmental problems, was "often presented as a disaster-in-waiting," but, Lomborg wrote, "these statements are often grossly exaggerated." He also wrote in an undocumented assertion that melting glaciers would be "a boon now."[10] However, six months after *Cool It* was published, the *Guardian* reported:

> The world's glaciers are melting faster than at any time since records began, threatening catastrophe for hundreds of millions of people and their eco-systems. The details are revealed in the latest report from the World Glacier Monitoring Service and will add to growing alarm about the rise in sea levels and increased instances of flooding, avalanches and drought. . . .
>
> Experts have been monitoring 30 glaciers around the world for nearly three decades and the most recent figures, for 2006, show the biggest ever "net loss" of ice. Achim Steiner, head of the UN Environment Programme (UNEP), told The Observer that melting glaciers were now the 'loudest and clearest' warning signal of global warming.

In a section that undermines Lomborg's unsupported assertion that melting glaciers will be a "boon now," the *Guardian* piece stated:

> The problem [of melting glaciers] could lead to failing infrastructure, mass migration and even conflict. "We're talking about something that happens in your and my lifespan. We're not talking about something hypothetical, we're talking about something dramatic in its consequences," he [Steiner] said. Lester Brown, of the influential US-based Earth Policy Institute, said the problem would have global ramifications, as farmers in China and India struggled to irrigate their crops. "This is the biggest predictable effect on food security in history as far as I know," said Brown.[11]

Similarly, in April 2008, seven months after *Cool It* was published, Reuters reported:

> Glaciers and mountain snow are melting earlier in the year than usual, meaning the water has already gone when millions of people need it during the summer when rainfall is lower, scientists warned on Monday. "This is just a time bomb," said hydrologist Carmen de Jong at a meeting of geoscientists in Vienna. Those areas most at risk from a lack of water for drinking and agriculture include parts of the Middle East, southern Africa, the United States, South America and the Mediterranean. Rising global temperatures mean the melt water is occurring earlier and faster in the year and the mountains may no longer be able

> to provide a vital stop gap. "In some areas where the glaciers are small they could be gone in 30 or 50 years time and a very reliable source of water, especially for the summer months, may be gone."
>
> De Jong was referring to parts of the Mediterranean where her research is focused but she said this threat also applies to the entire Alps region and other global mountain sources. Daniel Viviroli, from the University of Berne, believes nearly 40 percent of mountainous regions could be at risk, as they provide water to populations which cannot get it elsewhere. He says the earth's sub-tropic zones, which are home to 70 percent of the world's population, are the most vulnerable.[12]

Likewise, a little more than a year after *Cool It* was published, *Discovery News* reported: "Glaciers high in the Himalayas are dwindling faster than anyone thought, putting nearly a billion people living in South Asia in peril of losing their water supply. Throughout India, China, and Nepal, some 15,000 glaciers speckle the Tibetan Plateau, some of the highest land in the world. There, perched in thin, frigid air up to 7,200 meters (23,622 feet) above sea level, the ice might seem secluded from the effects of global warming. But just the opposite is proving true, according to new research published last week in the journal *Geophysical Research Letters.*"[13]

One of Lomborg's arguments in *Cool It* was that continental Antarctica had been cooling for decades despite global warming, that it is too cold for ice to begin to melt there, that it would accumulate rather than shed ice on a net basis this century, and that Antarctica would thus contribute a net decrease in sea levels. Lomborg also argued that the breakup of the Larsen-B ice shelf on the Antarctic Peninsula had more to do with natural climate variability than with man-made global warming.

As I have already pointed out, however, the sources that Lomborg cited failed to support his claim that "96 percent of Antarctica has cooled," and a *Nature* study published in January 2009 reported that "warming extends well beyond the Antarctic Peninsula to cover most of West Antarctica," that "West Antarctica warming exceeds 0.1°C per decade over the past 50 years," and that "the continent-wide average near-surface temperature trend is positive."[14] Regarding Lomborg's claims that most of Antarctica is too cold for ice to melt and that it would accumulate rather than shed ice this century overall, a February 2008 study in *Nature Geoscience* reported "a near-zero loss" of ice on the East Antarctic ice sheet in 1980–2004 but "widespread losses" in West Antarctica, with a loss increase of 59% of ice mass in the past ten years and 140% in the Antarctic Peninsula over the same period.[15] An article about the study in the *San Francisco Chronicle* reported: "Antarctica's massive coastal glaciers are quickly melting into the sea as the oceans around the continent grow warmer—and the pace of ice loss is speeding up. An international satellite network measuring the thickness of the glaciers as they shrink year by year has found that the glaciers have melted so rapidly during the past 10 years that the continent is losing almost as much ice as Greenland, according to researchers gathering the satellite data."[16]

Furthermore, not only was man-made global warming a major factor in the collapse of the Larsen-B ice shelf, contrary to what Lomborg argued, but the phenomenon of collapsing ice shelves on the Antarctic Peninsula is clearly migrating southward and thus inexorably closer to the continent. In March 2008, six months after *Cool It* was published, the Associated Press reported that "a chunk of ice seven times the size of Manhattan" from the Wilkins Ice Shelf on the southwest coast of the Antarctic Peninsula "suddenly collapsed"—"the result of global warming," said British Antarctic Survey scientist David Vaughan.[17] In July 2008, ten months after *Cool It* was published, the Australian Broadcasting Corporation reported: "It might be the depths of winter in Antarctica but scientists say they're seeing more signs of global warming on the frozen continent. New satellite photos show the Wilkins Ice Shelf is even closer to breaking from the peninsula. Experts say the effects of warming there now look irreversible. That may eventually lead to rising sea levels in more populated parts of the world."[18] And in January 2009, referring to the Wilkins Ice Shelf, Reuters reported: "A huge Antarctic ice shelf is on the brink of collapse with just a sliver of ice holding it in place, the latest victim of global warming that is altering maps of the frozen continent."[19]

Another of Lomborg's claims in *Cool It* was that sea levels would rise precisely one foot by the end of this century. Yet in April 2008, only seven months after *Cool It* was published, BBC News reported that "sea levels could rise by up to one-and-half metres [nearly 5 feet] by the end of this century, according to a new scientific analysis."[20] In December 2008, the *Guardian* reported that "sea level could rise by 150cm [nearly 5 feet], US scientists warn."[21] And in March 2009, the *San Francisco Chronicle* reported: "Driven by global warming, the ocean is expected to rise nearly 5 feet along California's coastline by the end of the century, hitting San Francisco Bay the hardest of all, according to a state study released Wednesday [March 11]."[22] The *Chronicle* continued: "Rising seas, storms and extreme high tides are expected to send saltwater into low-lying areas, flooding freeways, the Oakland and San Francisco airports, hospitals, power plants, schools and sewage plants. Thousands of structures at risk are the homes of low-and middle-income people, the study said. Vast wetlands that nourish fish and birds and act as a buffer against flooding will be inundated and could turn into dead pools. Constructing seawalls and levees, if needed, could cost $14 billion plus an annual maintenance cost of $1.4 billion, the study said."

Also in March 2009, the *Guardian* reported:

> Scientists will warn this week that rising sea levels, triggered by global warming, pose a far greater danger to the planet than previously estimated. There is now a major risk that many coastal areas around the world will be inundated by the end of the century because Antarctic and Greenland ice sheets are melting faster than previously estimated. (With much of the country already below sea level, even a small rise would be devastating for the Dutch.) Low-lying areas including Bangladesh, Florida, the Maldives and the Netherlands face catastrophic flooding, while, in Britain, large areas of the Norfolk Broads and the Thames estuary are likely to disappear by 2100. In addition, cities including London, Hull and Portsmouth will need new flood defences.

> "It is now clear that there are going to be massive flooding disasters around the globe," said Dr David Vaughan, of the British Antarctic Survey. "Populations are shifting to the coast, which means that more and more people are going to be threatened by sea-level rises."[23]

Also in March 2009, in an article titled, "Northeast US To Suffer Most from Future Sea Rise," the Associated Press reported:

> The northeastern U.S. coast is likely to see the world's biggest sea level rise from man-made global warming, a new study predicts. However much the oceans rise by the end of the century, add an extra 8 inches or so for New York, Boston and other spots along the coast from the mid-Atlantic to New England. That's because of predicted changes in ocean currents, according to a study based on computer models published online Sunday in the journal *Nature Geoscience.* An extra 8 inches—on top of a possible 2 or 3 feet of sea rise globally by 2100—is a big deal, especially when nor'easters and hurricanes hit, experts said.
>
> "It's not just waterfront homes and wetlands that are at stake here," said Donald Boesch, president of the University of Maryland Center for Environmental Science, who wasn't part of the study. "Those kind of rises in sea level when placed on top of the storm surges we see today, put in jeopardy lots of infrastructure, including the New York subway system."[24]

Though Lomborg sarcastically wrote in *Cool It* that "another of the most doom-laden impacts from global warming is the rising sea levels,"[25] and tied his itemization of a one-foot sea-level increase to figures in the 2007 IPCC assessment report that do not exist,[26] studies published both before and after the publication of *Cool It* pointed to a much higher increase in sea levels as a result of global warming.

Lomborg also argued in *Cool It* that "we will be able to feed the world ever better" in a warming world.[27] However, an FAO report issued in January 2009 projected that "global food production . . . must double by 2050 to head off mass hunger, the head of the UN's Food and Agriculture Organisation said on Monday." The report referred to a "food crisis [that] pushed another 40 million people into hunger in 2008," which increased the number "of undernourished people [worldwide] to 973 million."[28]

Though Lomborg devoted a sidebar article in *Cool It* about what he considered to be an exaggerated threat to Emperor penguins from global warming, the *Proceedings of the National Academy of Sciences* reported in January 2009, according to the *Independent*, that "the Emperor penguin is marching towards extinction because the Antarctic sea ice on which it depends for survival is shrinking at a faster rate than the bird is able to evolve if it is to avoid disaster." The *Independent* reported: "By the end of the century there could be just 400 breeding pairs of Emperor penguins left standing, a dramatic decline from the population [of] about 6,000 breeding pairs that existed in the 1960s, scientists estimated."[29]

Lomborg also argued that we should forgo prioritizing the reduction of greenhouse emissions as a coordinated global response to man-made climate change. Lomborg complained throughout *Cool It* about environmentalists who had called for cuts in greenhouse emissions. The response of the international community has been largely consistent with his complaint, and the failure to reduce greenhouse emissions has greatly contributed to dangerous man-made levels of atmospheric CO_2. In October 2007, one month after *Cool It* was published, Bloomberg reported that "the carbon dioxide concentration in the atmosphere is rising faster than expected as oceans and the land absorb a lower proportion of the gas and humans become less efficient at producing energy, an Australian-led team of scientists said."[30] In November 2007 Reuters reported: "Levels of carbon dioxide, the main greenhouse gas emitted by burning fossil fuels, hit a record high in the atmosphere in 2006, accelerating global warming, the World Meteorological Organization said."[31]

In September 2008, the Global Carbon Project said that CO_2 was increasing at a rate that exceeds even the IPCC's worst-case estimates: "Despite the increasing international sense of urgency, the growth rate of emissions continued to speed up, bringing the atmospheric CO_2 concentration to 383 parts per million (ppm) in 2007. Anthropogenic CO_2 emissions have been growing about four times faster since 2000 than during the previous decade, despite efforts to curb emissions in a number of Kyoto Protocol signatory countries."[32] In other words, from 2000 to 2007—the period during which Lomborg argued in *The Skeptical Environmentalist* and *Cool It* that global warming was no catastrophe and that there was no need to prioritize reductions in greenhouse emissions—CO_2 emissions grew "about four times faster" than in the previous decade, "above even the most fossil-fuel intensive scenario of the Intergovernmental Panel on Climate Change,"[33] as the Global Carbon Project reported.

In March 2009 the *Independent* reported: "The world's best efforts at combating climate change are likely to offer no more than a 50–50 chance of keeping temperature rises below the threshold of disaster, according to research from the UK Met Office. The key aim of holding the expected increase to 2C, beyond which damage to the natural world and to human society is likely to be catastrophic, is far from assured, the research suggests, even if all countries engage forthwith in a radical and enormous crash programme to slash greenhouse gas emissions—something which itself is by no means guaranteed." The *Independent* report explained further that "the Hadley Centre's simulation indicates that even if global emissions of carbon dioxide, the main greenhouse gas causing the warming, were to be slashed at a very high rate the chances of holding the [temperature] rise" at the 2°C threshold "are no better than even."[34]

Since 2000, even as Lomborg emphasized his opposition to serious reductions of greenhouse emissions, scientists were looking at "the consequences of CO_2 buildup beyond present-day concentrations of 385 parts per million, and then completely stopping emissions after the peak." A study published in the January 2009 *Proceedings of the National Academy of Sciences* concluded that climate change is "largely irreversible" for the next thousand years even if carbon dioxide emissions could be abruptly halted.[35] In reporting on this study, led by Susan Solomon, director of the National Oceanic and Atmospheric Administration, the Associated Press reported:

> Before the industrial revolution the air contained about 280 parts per million of carbon dioxide. That has risen to 385 ppm today, and politicians and scientists have debated at what level it could be stabilized.

Solomon's paper concludes that if CO_2 is allowed to peak at 450–600 parts per million, the results would include persistent decreases in dry-season rainfall that are comparable to the 1930s North American Dust Bowl in zones including southern Europe, northern Africa, southwestern North America, southern Africa and western Australia.

Gerald Meehl, a senior scientist at the National Center for Atmospheric Research, said, "The real concern is that the longer we wait to do something, the higher the level of irreversible climate change to which we'll have to adapt."[36]

Although Lomborg has not been an official decision-maker throughout the past decade, he has been influential in supporting government inaction on greenhouse emissions.

In a 2008 paper published in *The Open Atmospheric Science Journal*, James Hansen and colleagues similarly addressed the issue of atmospheric CO_2: "If humanity wishes to preserve a planet similar to that on which civilization developed and to which life on Earth is adapted, paleoclimate evidence and ongoing climate change suggest that CO_2 will need to be reduced from its current 385 ppm to at most 350 ppm. . . . If the present overshoot of this target CO_2 [of 350 ppm] is not brief, there is a possibility of seeding irreversible catastrophic effects."[37] Likewise, in a speech at the UN climate summit in Poland in December 2008, according to the Web site Grist, "Al Gore argued that older targets for reducing global-warming pollution are out of date, and that world leaders should aim to reduce the amount of carbon dioxide in the atmosphere to 350 parts per million. 'Even a goal of 450 parts per million, which seems so difficult today, is inadequate,' said Gore."[38]

If the evidence in the Solomon study and the Hansen paper above is even roughly accurate, then the relatively broad tolerance of and support for Lomborg's Theorem and Lomborg's Corollary contributed to an enormously damaging delay in the reduction of greenhouse emissions. It would also indicate that the urgent messages from Hansen and Gore to focus on reducing greenhouse emissions were scientifically sound, in contrast to Lomborg's claims and recommendations.

The evidence would also indicate that the Worldwatch Institute's assessment of global warming—whose annual State of the World reports Lomborg parodied in the subtitle of his 2001 book, *The Skeptical Environmentalist: Measuring the Real State of the World*—was also right. A Worldwatch Institute's press release, dated January 13, 2009, on its 2009 *State of the World* report, began: "The world will have to reduce emissions more drastically than has been widely predicted, essentially ending the emission of carbon dioxide by 2050 to avoid catastrophic disruption to the world's climate."[39] The report itself,[40] like each of the preceding *State of the World* reports, is an excellent example of environmentalism combining scientific expertise, intellectual integrity, and social responsibility.

An appropriate way to end this book is to present two views—both published on the same day in March 2009—of where we are and what we face as a civilization with respect to global warming. One view is described by Michael McCarthy, a science and environmental journalist at the *Independent* who was reporting on the comments of Nicholas Stern, the British economist "who produced the single most influential political document on climate change"—the 2006 Stern Review.[41] Reporting that Stern "says he underestimated the risks of global warming and the damage that could result from it," McCarthy wrote:

> The situation was worse than he had thought when he completed his review two-and-a-half years ago, he told a conference [of climate scientists in Copenhagen] yesterday, but politicians do not yet grasp the scale of the dangers now becoming apparent.
>
> "Do politicians understand just how difficult it could be, just how devastating rises of 4C, 5C or 6C could be? I think, not yet," Lord Stern posed to the meeting of scientists in Copenhagen. "A rise of 5C [9°F] would be a temperature the world has not seen for 30 to 50 million years. We've been around only 100,000 years as human beings. We don't know what that's like. We haven't seen 3C [5.4°F] for a few million years, and we don't know what that looks like either."
>
> Lord Stern said new research done in the past two or three years had

> made it clear there were "severe risks" if global temperature rose by the predicted 4C to 7C [7.2 to 12.6°F] by 2100. Agriculture would be destroyed and life would be impossible over much of the planet, the former World Bank chief economist said.[42]

On the same day that McCarthy's piece appeared, a Gallup poll that largely reflected Lomborg's position on global warming was issued. Agence France Presse summarized the results: "More Americans than at any time in the past decade believe that the seriousness of global warming is being exaggerated, a Gallup poll showed Thursday. Forty-one percent of Americans told Gallup pollsters that they are doubtful that global warming is as serious as the mainstream media are reporting, putting public skepticism about the hot-button issue at the highest level recorded by Gallup in more than a decade."[43]

It would be difficult to ascertain with certainty to what extent the 41 percent was influenced by Lomborg and his supporters in the U.S. news media. Nevertheless, *Time* magazine presumably included Lomborg in its list of the one hundred most influential persons in 2004 for a reason. And it would be difficult to argue that Lomborg's favorable press coverage from 2001 to 2009 had an insignificant impact on the public's views about global warming. Indeed, one could credibly argue that Lomborg's ultimate legacy will be that he helped leave nearly half the U.S. public unconvinced that global warming is a serious threat to civilization, when the U.S. Constitution requires a minimum of two-thirds (sixty-seven of one hundred) of the U.S. Senate to ratify a potentially decisive December 2009 climate treaty in Copenhagen.

One can convincingly maintain that Lomborg's books are an assault on science, as *Scientific American* did when it convened its forum of distinguished scientists to write an important early rebuttal—"Science Defends Itself Against The Skeptical Environmentalist"—to Lomborg. One might also argue that the success of Lomborg's books in a cultural sense was one manifestation of a broader "assault on reason," an apt coinage of our current overall predicament that Al

Gore chose as the title of his recent book,[44] which describes a systemic breakdown of rational consideration of the major challenges facing the United States and the world. That the twenty-year assessment of climate science by the IPCC's scientists would be compared unfavorably with Lomborg's *The Skeptical Environmentalist* and *Cool It* in so many influential circles in the United States—given the evident problems with Lomborg's scholarship—is at least one inescapable affirmation of Gore's thesis.

Beyond these final comments, it is difficult to label *Cool It* and the larger Lomborg phenomenon. At this point we can but issue an appeal to Lomborg's prospective book publishers to check—*before* publication—whether his sources support his assertions or not. Doing so could save the world. Or at least the one we now face.

NOTES

Chapter 1. 2001: A Theorem's Odyssey

1. Bjørn Lomborg, *The Skeptical Environmentalist: Measuring the Real State of the World* (Cambridge, U.K.: Cambridge University Press, 2001), 4.
2. Ibid.
3. Ibid.
4. See Bjørn Lomborg's Web site at http://www.lomborg.com/.
5. Bjørn Lomborg, *Cool It: The Skeptical Environmentalist's Guide to Global Warming* (New York: Knopf, 2007), ix.
6. Ibid., 53, 3–112.
7. Lomborg, *The Skeptical Environmentalist*, xix.
8. Lomborg, *The Skeptical Environmentalist*, publisher's note, "The Skeptical Environmentalist," inside front jacket flap.
9. Stephen Schneider, "Global Warming: Neglecting the Complexities," *Scientific American*, vol. 286, no. 1, January 2002, 65.
10. "Scientist at Work: Bjorn Lomborg; From an Unlikely Quarter, Eco-Optimism," *New York Times*, August 7, 2001.
11. "Greener Than You Think: 'The Skeptical Environmentalist: Measuring the Real State of the World' by Bjorn Lomborg," *Washington Post*, October 21, 2001.
12. "Why All Those Dire Predictions Have No Future," *Wall Street Journal*, October 2, 2001.
13. Grist: Environmental News and Commentary, at http://www.grist.org.
14. Other contributors to the Grist Forum, "Something Is Rotten in the State of Denmark: A Skeptical Look at The Skeptical Environmentalist," were: Emily Matthews, a forest expert and senior associate with the World Resources Institute; Al Hammond, a senior scientist at World Resources Institute; Devra Davis, an epidemiologist and environmental health researcher; David Nemtzow, president of the Alliance to Save Energy; and Kathryn Schulz, associate editor at Grist. The Grist forum on Lomborg's *The Skeptical Environmentalist* was last accessed on July 22, 2009, at http://www.grist.org/advice/books/2001/12/12/of/.
15. Lester Brown, "Bjorn Again: On Bjorn Lomborg and Population," Grist,

December 12, 2001, last accessed on July 22, 2009, at http://www.grist.org/article/bjorn/.

16. Ibid.
17. Edward O. Wilson, "Vanishing Point: On Bjorn Lomborg and Extinction," Grist, December 12, 2001, last accessed on July 22, 2009, at http://www.grist.org/article/point/.
18. See Norman Myers and Julian L. Simon, *Scarcity or Abundance? A Debate on the Environment* (New York: W.W. Norton, 1994).
19. Norman Myers, "Specious: On Bjorn Lomborg and Species Diversity," Grist, December 12, 2001, last accessed on July 22, 2009, at http://www.grist.org/article/specious/.
20. Ibid.
21. Stephen Schneider, "Hostile Climate: On Bjorn Lomborg and Climate Change," Grist, December 12, 2001, last accessed on July 22, 2009, at http://www.grist.org/article/hostile/.
22. Ibid.
23. "Background: UCS Examines 'The Skeptical Environmentalist,' " Union of Concerned Scientists, last accessed on July 22, 2009, at http://www.ucsusa.org/global_warming/science_and_impacts/global_warming_contrarians/ucs-examines-the-skeptical.html.
24. Ibid.
25. John Rennie, "Misleading Math About the Earth: Science Defends Itself Against The Skeptical Environmentalist," *Scientific American*, vol. 286, no. 1, January 2002, 61; commentary by Schneider, Holdren, Bongaarts, and Lovejoy appeared at 62–72.
26. Schneider, "Global Warming: Neglecting the Complexities," 63.
27. See Lomborg, *The Skeptical Environmentalist*, Figure 151, 285–86.
28. Schneider, "Global Warming: Neglecting the Complexities," 63–65.
29. John P. Holdren, "Energy: Asking the Wrong Question," *Scientific American*, vol. 286, no. 1, January 2002, 65.
30. John Bongaarts, "Population: Ignoring Its Impact," *Scientific American*," vol. 286, no. 1, January 2002, 67–69.
31. Ibid.
32. Thomas Lovejoy, "Biodiversity: Dismissing Scientific Process," *Scientific American*, vol. 286, no. 1, January 2002, 69–71.
33. Lomborg, *The Skeptical Environmentalist*, 253–54; Lovejoy, "Biodiversity: Dismissing Scientific Process," 70, 71.
34. Lovejoy, "Biodiversity: Dismissing Scientific Process," 70, 71.
35. Lomborg, *The Skeptical Environmentalist*, 110.

36. Ibid., 116.
37. Lovejoy, "Biodiversity: Dismissing Scientific Process," 71.
38. Nigel Dudley, Jean-Paul Jeanrenaud, and Sue Stolton, "The Year the World Caught Fire: A Report to WWF International," 1997, last accessed on July 23, 2009, at http://www.equilibriumresearch.com/upload/document/theyeartheworldcaughtfire.pdf.
39. Ibid.
40. Lomborg, *The Skeptical Environmentalist*, 116, 377 (note 835). Lomborg also referenced an article in the *Christian Science Monitor*, in which Goldammer, of the Max Planck Institute for Chemistry in Germany, stated, "There is no indication at all that 1997 was an extraordinary fire year for Indonesia or the world at large." However, other experts also were quoted in the article. Nigel Dudley of World Wildlife Fund stated: "This year's burning was very serious and will probably turn out to be the worst on record." Don Henry, director of the World Wildlife Fund's Global Forest Program, stated: "The fires are a symptom, not a cause of deforestation. We're seeing tropical forests vanishing at an alarming scale due to short-sighted land-use policies. Fire is just one of the means by which forests are being cleared." Stephen Pyne, a scientist at Arizona State University, stated: "This isn't a problem of too much fire. It's a political and land-use problem. Planned fires are being set in places and ways that are inappropriate." The *Monitor*'s Colin Woodard reported: "The burning of forests releases huge amounts of carbon into the atmosphere, which may contribute to global warming. If the forests aren't able to regrow—because they've been replaced with pasture or crops—the land will not absorb as much carbon from the atmosphere. In a six-month period, the Indonesian fires [of 1997] may have released more carbon dioxide than all the cars and power stations in Western Europe release in a year, estimates University of Nottingham (England) peat specialist Jack Rieley." Woodard also reported that "last year may have been the worst yet for the destruction of the Amazon rain forests," and that "satellites detected 20,469 fires [in the Amazon rain forest] last fall, up 28 percent from 1996." With respect to Goldammer's personal communication with Lomborg and the article in the *Christian Science Monitor*, Lomborg reported only that "Goldhammer said that 'there is no indication at all that 1997 was an extraordinary fire year for Indonesia or the world at large.' " See "Lessons from 'the Year the Earth Caught Fire,' " *Christian Science Monitor*, February 4, 1998, last accessed on July 23, 2009, at http://www.csmonitor.com/1998/0204/020498.us.us.4.html.

41. Emily Matthews, "Not Seeing the Forest for the Trees: On Bjorn Lomborg and Deforestation," Grist, December 12, 2001, at http://www.grist.org/article/for/. Like Lovejoy, Matthews examined Lomborg's analysis in *The Skeptical Environmentalist* of the 1997–98 fires in Indonesia. She commented:

> Lomborg devotes an entire page to Indonesia's fires of 1997–1998, acknowledging that they were serious but also claiming that they were not out of the ordinary. He criticizes WWF for estimating that 2 million hectares burned and contrasts this claim with the "official Indonesian estimate" of 165,000–219,000 hectares. He notes that the WWF estimate included both forest and non-forestland, but does not point out that the official Indonesian estimate he quotes was for forestland only. He then claims, citing a 1999 United Nations Environment Programme report, that subsequent "satellite-aided counting" indicated that upwards of 1.3 million hectares of forest and timberlands may have burned.
>
> The official Indonesian estimate of 520,000 burned hectares of forest and non-forest land was based on reports by plantation owners—who were responsible for much of the deliberate fire-setting and had no incentive to report accurately. This estimate was quickly challenged by the German-supported Integrated Forest Fires Management Project, which, using satellite data and ground checks, produced convincing evidence that fires had actually burned some 5.2 million hectares in 1998 alone—10 times the Indonesian government's estimate. Informed of this data gap, the Indonesian Ministry of Forestry and Estate Crops effectively instructed the governor of East Kalimantan (the province that suffered the worst fires) not to allow the new data to be made public, citing the need "to protect national stability." Despite strong official protests from the German government, the Indonesians never retracted their original estimate or made the new data public.
>
> Regarding estimates of how much forest actually burned, Lomborg cites a UNEP report, which in turn refers to an analysis, "A Study of the 1997 Fires in Southeast Asia Using SPOT Quicklook Mosaics," that was based on 766 satellite images. These images covered the islands of Kalimantan and Sumatra only, for just August to December 1997. The study did not examine burn areas for 1998, nor did it take into account fires on other islands. The UNEP report

states that this estimate represents "only a lower limit estimate of the area burned," although Lomborg's readers are not so informed.

An analysis by the Singapore Centre of Remote Imaging, Sensing, and Processing using the same satellite images yielded a total burn area estimate for 1997 and 1998 of nearly 8 million hectares. In 1999, a technical team funded by the Asian Development Bank and working through the Indonesian National Development Planning Agency aggregated and analyzed all available data sources and estimated that the area burned during 1997–98 totaled more than 9.7 million hectares, of which some 4.6 million hectares were forest.

Thus, the most authoritative consensus estimate of the extent of forests burned during the Indonesian fires of 1997–98 is more than twice the WWF estimate that is derided by Lomborg.

42. Howard Friel and Richard Falk, *The Record of the Paper: How the New York Times Misreports US Foreign Policy* (London: Verso, 2004); Friel and Falk, *Israel-Palestine on Record: How the New York Times Misreports Conflict in the Middle East* (New York: Verso, 2007).
43. "Summary for Policymakers," *Climate Change 2007: Synthesis Report*, November 17, 2007, 1–23, last accessed on July 23, 2009, at http://www.ipcc.ch/pdf/assessment-report/ar4/syr/ar4_syr_spm.pdf.
44. "Nobel Lecture: Al Gore: The Nobel Peace Prize 2007," Nobelprize.org, last accessed on July 23, 2009, at http://nobelprize.org/nobel_prizes/peace/laureates/2007/gore-lecture_en.html; "Nobel Lecture: Intergovernmental Panel on Climate Change: The Nobel Peace Prize 2007," Nobelprize.org, last accessed on July 23, 2009, at http://nobelprize.org/nobel_prizes/peace/laureates/2007/ipcc-lecture.html.
45. "Bert Bolin, 82, Is Dead; Led U.N. Climate Panel," *New York Times*, January 4, 2008.
46. Elizabeth Kolbert, *Field Notes from a Catastrophe: Man, Nature, and Climate Change* (New York: Bloomsbury, 2006).
47. James Gustave Speth, *Red Sky at Morning: America and the Crisis of the Global Environment* (New Haven: Yale University Press, 2004).
48. Spencer Weart, *The Discovery of Global Warming* (Cambridge: Harvard University Press, 2003).
49. "Bert Bolin, 82, Is Dead."
50. Lomborg, *Cool It*, 8.
51. Weart, *The Discovery of Global Warming*, 71.
52. These included: Jule G. Charney, Massachusetts Institute of Technology,

chair; Akio Arakawa, University of California, Los Angeles; D. James Baker, University of Washington; Bert Bolin, University of Stockholm; Robert E. Dickinson, U.S. National Center for Atmospheric Research; Richard M. Goody, Harvard University; Cecil E. Leith, National Center for Atmospheric Research; Henry M. Strommel, Woods Hole Oceanographic Institution; and Carl I. Wunsch, Massachusetts Institute of Technology.

53. "Carbon Dioxide and Climate: A Scientific Assessment: Report of an Ad Hoc Study Group on Carbon Dioxide and Climate," Climate Research Board, Assembly of Mathematical and Physical Sciences, National Research Council, National Academy of Sciences, Washington, D.C., July 23–27, 1979, vii, viii, last accessed on July 23, 2009, at http://www.atmos.ucla.edu/brianpm/download/charney_report.pdf.
54. Ibid., ii.
55. "Carbon Dioxide Accumulation in the Atmosphere, Synthetic Fuels and Energy Policy," Committee on Governmental Affairs, United States Senate, U.S. Government Printing Office, July 30, 1979, 13, 20.
56. "Study Finds Warming Trend That Could Raise Sea Levels," *New York Times*, August 22, 1981.
57. "Global Warming Has Begun, Expert Tells Senate," *New York Times*, June 24, 1988.
58. Intergovernmental Panel on Climate Change, *Climate Change: The IPCC Scientific Assessment* (Cambridge, UK: Cambridge University Press, 1990), iii.
59. Ibid., xi.
60. Ibid., xxii.
61. Lomborg, *Cool It*, 8.
62. Intergovernmental Panel on Climate Change, *Climate Change: The IPCC Scientific Assessment* (1990), xxiii.
63. National Academy of Sciences, *Policy Implications of Greenhouse Warming* (Washington, D.C.: National Academy Press, 1991), 2.
64. "Summary for Policymakers," *Climate Change 2007: Synthesis Report*, 8.

Chapter 2. On Polar Bears

1. Bjørn Lomborg, *Cool It: The Skeptical Environmentalist's Guide to Global Warming* (New York: Knopf, 2007), 4.
2. "Special Report: Global Warming: Be Worried; Be Very Worried," *Time*, April 3, 2006, last accessed on July 23, 2009, at http://www.time.com/time/covers/0,16641,20060403,00.html. Click on "Read the Cover Story."

3. Lomborg, *Cool It*, 4.
4. Ibid., 5.
5. Ibid., 5–6.
6. "Polar Bears: Proceedings of the 13th Working Meeting of the IUCN/SSC Polar Bear Specialist Working Group, 23–28 June 2001, Nuuk, Greenland," International Union for the Conservation of Nature, 2001, Table 1, 22, last accessed on July 23, 2009, at http://www.carnivoreconservation.org/files/actionplans/pbsg13proc.pdf.
7. Ibid.
8. Ibid.
9. Ibid., 27.
10. Ibid., 29.
11. Lomborg, *Cool It*, 5.
12. Ibid., 167–68.
13. *Arctic Climate Impact Assessment Scientific Report* (Cambridge, U.K.: Cambridge University Press, 2004), available at http://www.acia.uaf.edu/pages/scientific.html. The three most detailed sections on polar bears in the ACIA are at pp. 487, 509, and 632.
14. Ibid., 487.
15. Ibid., 509.
16. Lomborg, *Cool It*, 6–7.
17. To compare what Lomborg wrote in *Cool It* to what the ACIA reported in this regard, see Lomborg, *Cool It*, 6–7, 168, and *Arctic Climate Impact Assessment*, 2004, 509.
18. Lomborg, *Cool It*, 5.
19. "Polar Bears: Proceedings of the 13th Working Meeting of the IUCN/SSC Polar Bear Specialist Working Group," 27.
20. *Arctic Climate Impact Assessment*, 2004, 504.
21. "Polar Disasters: More Predictable Distortions of Science," Cato Institute, November 22, 2004, last accessed on July 23, 2009, at http://www.cato.org/pub_display.php?pub_id=2888.
22. Ibid.
23. "Polar Bears: Proceedings of the 13th Working Meeting of the IUCN/SSC Polar Bear Specialist Working Group," 28.
24. Rajumund Przybylak, "Temporal and Spatial Variation of Surface Air Temperature over the Period of Instrumental Observations in the Arctic," *International Journal of Climatology*, vol. 20, no. 6, May 2000, 587–614.
25. Lomborg, *Cool It*, 5. Lomborg's statement, which he referenced to the Przybylak paper, reads in full: "Contrary to what you might expect—and

what was not pointed out in any of the recent stories—the two populations in decline come from areas where it has actually been getting colder over the past fifty years, whereas the two increasing populations reside in areas where it is getting warmer."

26. Ibid., 6.
27. Ian Stirling, Nicholas J. Lunn, and John Iacozza, "Long-Term Trends in the Population Ecology of Polar Bears in Western Hudson Bay in Relation to Climatic Change," *Arctic*, vol. 52, no. 3, September 1999, 302.
28. Ibid., 294.
29. Steven C. Amstrup, George M. Durner, Geoff York, et al., "Polar Bears: Sentinel Species for Climate Change," United States Geological Survey: Alaska Science Center, 2006, Slide 44.
30. Jon Aars, Nicholas J. Lunn, and Andrew E. Derocher, eds., "Polar Bears: Proceedings of the 14th Working Meeting of the IUCN/SSC Polar Bear Specialist Group, 20–24 June 2005, Seattle, Washington, USA: Occasional Paper of the IUCN Species Survival Commission No. 32," 57–58, last accessed on July 24, 2009, at http://pbsg.npolar.no/export/sites/pbsg/en/docs/PBSG14proc.pdf.
31. "Polar Bears: Proceedings of the 13th Working Meeting of the IUCN/SSC Polar Bear Specialist Working Group," 54.
32. Eric V. Regehr, Nicholas J. Lunn, Steven C. Amstrup, and Ian Stirling, "Effects of Earlier Sea-Ice Breakup on Survival and Population Size of Polar Bears in Western Hudson Bay," *Journal of Wildlife Management*, vol. 71, no. 8, November 2007, 2673.
33. Lomborg, *Cool It*, 6.
34. Ibid., 7–8.
35. "Polar Bears: Proceedings of the 13th Working Meeting of the IUCN/SSC Polar Bear Specialist Working Group," 22.
36. Lomborg, *Cool It*, 6, 168, 237.
37. Lomborg, *Cool It*, 168, 237; "Silly to Predict Their Demise," *Toronto Star*, May 1, 2006.
38. "Arctic Ice Melting Much Faster Than Thought," *Globe and Mail*, November 28, 2002.
39. "Arctic Ice Is Melting at Record Level, Scientists Say," *New York Times*, December 8, 2002.
40. "Climate Change Blamed as Largest Arctic Ice Shelf Breaks in Two After 3,000 Years," *Independent*, September 24, 2003.
41. "Arctic Ice Melting at Worrying Rate: NASA," Agence France Presse, October 24, 2003.

42. "Study Says Polar Bears Could Face Extinction," *Washington Post*, November 9, 2004.
43. Lomborg, *Cool It*, 7.
44. Ibid., 6.
45. "Fears Over Climate as Arctic Ice Melts at Record Level," *Guardian*, September 29, 2005.
46. "Even in Winter, Arctic Ice Melting," *San Francisco Chronicle*, September 14, 2006.
47. Al Gore, *An Inconvenient Truth: The Planetary Emergency of Global Warming and What We Can Do About It* (New York: Rodale, 2006), 146.
48. Lomborg, *Cool It*, 5–6.
49. Charles Monnett, Jeffrey S. Gleason, and Lisa M. Rotterman, "Potential Effects of Diminished Sea Ice on Open-Water Swimming, Mortality, and Distribution of Polar Bears During Fall in the Alaskan Beaufort Sea," Minerals Management Service, Alaska OCS Region, Environmental Studies Section, Environmental Assessment Section, 2005, last accessed on July 24, 2009, at http://www.mms.gov/alaska/ess/Poster%20Presentations/MarineMammalConference-Dec2005.pdf.
50. Ibid.
51. "Polar Bears Drown as Ice Shelf Melts," *Sunday Times* (London), December 18, 2005.
52. Lomborg, *Cool It*, 6.
53. Ibid., 7.
54. Ibid.
55. *Arctic Climate Impact Assessment*, 2004, 997–98.
56. Ibid., 997–99.
57. Lomborg, *Cool It*, 7.
58. *Arctic Climate Impact Assessment*, 2004, 256.
59. Lomborg, *Cool It*, 7, 8, 9.
60. Ibid., 8.

Chapter 3. On Lomborg's Endnotes

1. Bjørn Lomborg, *The Skeptical Environmentalist: Measuring the Real State of the World* (Cambridge, U.K.: Cambridge University Press, 2001), 3.
2. Ibid.
3. Ibid., 4.
4. Ibid., 3–4, 353.
5. Ibid., 4, 354.

6. Ibid., 4.
7. See, for example: Herman E. Daly, *Beyond Growth: The Economics of Sustainable Development* (Boston: Beacon, 1997); Herman E. Daly and Joshua Farley, *Ecological Economics: Principles and Applications* (Washington, D.C.: Island, 2003); Bill McKibben, *Deep Economy: The Wealth of Communities and the Durable Future* (New York: Holt Paperbacks, 2008).
8. Lomborg, *The Skeptical Environmentalist*, 4, 354.
9. Ibid., 5.
10. Ibid.
11. "World Food Summit: November 13–17, 1996: Rome, Italy: Technical Background Document: FAO, 1966," Table 3, last accessed on July 24, 2009, at http://www.fao.org/docrep/003/w2612e/w2612e01.htm.
12. "The State of Food Insecurity in the World: Food Insecurity: When People Must Live with Hunger and Fear Starvation," FAO, 1999, Table 1, p. 29, last accessed on July 24, 2009, at ftp://ftp.fao.org/docrep/fao/007/x3114e/x3114e00.pdf.
13. Ibid., 4.
14. Ibid.
15. Ibid., 26.
16. "The State of Food Insecurity in the World, 2004: Monitoring Progress Toward the World Food Summit and Millennium Development Goals: Undernourishment Around the World: Counting the Hungry: Latest Estimates," 6, last accessed on July 24, 2009, at ftp://ftp.fao.org/docrep/fao/007/y5650e/y5650e00.pdf.
17. Lomborg, *The Skeptical Environmentalist*, 5.
18. Ibid.
19. Ibid.
20. Ibid., 5–6, 354.
21. Ibid., 6.
22. "Compendium of Statistics on Illiteracy — 1990 Edition," UNESCO, last accessed on July 24, 2009, at http://unesdoc.unesco.org/images/0008/000874/087413mb.pdf.
23. Lomborg, *The Skeptical Environmentalist*, Figure 5, p. 22.
24. Ibid., 81.
25. Ibid.; see endnotes 572 and 573, pp. 369–70, and the bibliographic reference, p. 496.
26. "UNESCO: Gender-Sensitive Education Statistics and Indicators: A Practical Guide," July 1997, Table 2, p. 10, last accessed on July 25, 2009, at http://unesdoc.unesco.org/images/0010/001091/109177e0.pdf.
27. Ibid.

28. Lomborg, *The Skeptical Environmentalist*, 6.
29. "World Food Summit: 13–17 November 1996, Rome, Italy: FAO Technical Background Document," FAO, 1996.
30. FAO, "The State of Food Insecurity in the World: 1999."
31. "World Food Summit: 13–17 November 1996, Rome, Italy: FAO Technical Background Document," Table 3; "The State of Food Insecurity in the World, 1999," Table 1, p. 29.
32. FAO, "The State of Food Insecurity in the World, 1999," Table 1, p. 29.
33. Ibid., 8.
34. "World Food Summit: 13–17 November 1996, Rome, Italy: FAO Technical Background Document," Table 3.
35. FAO, "The State of Food Insecurity in the World, 1999," 9.
36. FAO, "The State of Food Insecurity in the World, 2006," 23, last accessed on July 25, 2009, at ftp://ftp.fao.org/docrep/fao/009/a0750e/a0750e00.pdf.
37. "World Food Summit: 13–17 November 1996, Rome, Italy: FAO Technical Background Document," Table 3.
38. FAO, "The State of Food Insecurity in the World, 1999," 11.
39. Ibid., Table 1, pp. 29–30.
40. Lomborg, *The Skeptical Environmentalist*, 6–7.
41. United Nations Environment Program, "Global Environmental Outlook 2000," chapter 2: "The State of the Environment—Africa," last accessed on July 25, 2009, at http://www.unep.org/geo2000/english/0053.htm.
42. Lomborg, *The Skeptical Environmentalist*, 7.
43. "World Agriculture: Towards 2010: An FAO Study," 1995, last accessed on July 25, 2009, at http://www.fao.org/docrep/v4200e/v4200E00.htm; "World Food Prospect: Critical Issues for the Early Twenty-First Century," Food Policy Report, International Food Policy Research Institute, Washington, D.C., 1999, last accessed on July 25, 2009, at http://www.ifpri.org/pubs/fpr/fpr29.pdf.
44. "World Food Prospect: Critical Issues for the Early Twenty-First Century," 25.
45. Ibid., 6.
46. Ibid., Table 2, p. 10.
47. See Lomborg, *The Skeptical Environmentalist*, 156, 176, 289, 290, 291.
48. "Roads Not Taken," *New York Times*, April 25, 2003.
49. Lomborg, *The Skeptical Environmentalist*, 7.
50. Lomborg's endnote 27 references "FAO 2000a." The bibliographical reference for this citation reads in full: "FAO 2000a. Database, accessed in 2000: http://apps.fao.org/."

51. Lomborg, *The Skeptical Environmentalist*, 7.
52. Ibid. In his Notes and Bibliography, Lomborg cites John Boardman, "An Average Soil Erosion Rate For Europe: Myth or Reality," *Journal of Soil and Water Conservation*," vol. 53, no. 1, 1998, 46–50.
53. David Pimentel, C. Harvey, P. Resosudarmo, et al., "Environmental and Economic Costs of Soil Erosion and Conservation Benefits," *Science*, vol. 267, no. 5201, February 24, 1995, 1117–23.
54. Lomborg, *The Skeptical Environmentalist*, 7.
55. Ibid.
56. Pimentel, Harvey, Resosudarmo, et al., "Environmental and Economic Costs of Soil Erosion and Conservation Benefits," 1117.
57. Ibid.
58. Ibid.

Chapter 4. Global Warming Is "No Catastrophe"

1. Bjørn Lomborg, *Cool It: The Skeptical Environmentalist's Guide to Global Warming* (New York: Knopf, 2007), 53, 3–112.
2. IPCC Fourth Assessment Report (2007): Working Group II: Impacts, Adaptation, and Vulnerability, "Summary for Policymakers," 13–15, last accessed on July 25, 2009, at http://www.ipcc.ch/pdf/assessment-report/ar4/wg2/ar4-wg2-spm.pdf.
3. Lomborg referenced the IPCC on five other occasions in *Cool It*, chapter 2, that were relevant to his cost-benefit assessment of reducing greenhouse emissions. See "Notes," 173 (one reference to the IPCC), 174 (three references), 177 (one reference).
4. Lomborg, *Cool It*, 10–11.
5. Ibid., 11.
6. See Lomborg's sources for this summary in *Cool It*, 10–11, 169.
7. Lomborg, *Cool It*, 11.
8. Ibid.
9. Ibid., 169.
10. IPCC Fourth Assessment Report (2007): Working Group I: The Physical Science Basis, "Summary for Policymakers," 14 (Figure SPM.5), last accessed on July 25, 2009, at http://www.ipcc.ch/pdf/assessment-report/ar4/wg1/ar4-wg1-spm.pdf.
11. Lomborg, *Cool It*, 169. Lomborg cited: "IPCC, 2007b: fig. 10.3.1." Given Lomborg's citation system, this citation refers to the IPCC Fourth Assessment Report (2007): Working Group I: The Physical Science Basis.

However, there is no Figure 10.3.1 in the IPCC Fourth Assessment Report (2007), Working Group I. Nor is there a Figure 10.3.1 in the Working Group II or Working Group III sections.

12. "SRES" refers to Nebojsa Nakicenovic and Rob Swart, eds., *IPCC Special Report on Emissions Scenarios, 2000,* last accessed on July 25, 2009, at http://www.ipcc.ch/ipccreports/sres/emission/index.htm.
13. Nebojsa Nakicenovic and Rob Swart, eds., *IPCC Special Report on Emissions Scenarios, 2000,* "Summary for Policymakers," 3–4, last accessed on July 25, 2009, at http://www.ipcc.ch/pdf/special-reports/spm/sres-en.pdf.
14. Ibid., 3.
15. IPCC Fourth Assessment Report (2007): Working Group I: The Physical Science Basis, "Summary for Policymakers," 14 (Figure SPM.5).
16. Ibid.
17. Ibid., 13 (Table SPM.3).
18. Lomborg, *Cool It,* 13, 169.
19. Ibid.
20. Ibid., 13–14, 169.
21. Ibid., 14, 169.
22. Ibid., 14, 170.
23. Ibid., 14.
24. Ibid.
25. Ibid., 15.
26. See Lomborg, *Cool It,* 14–15, 170. The five studies referenced are: Kristie L. Ebi, David M. Mills, Joel B. Smith, and Anne Grambsch, "Climate Change and Human Health Impacts in the United States: An Update on the Results of the U.S. National Assessment," *Environmental Health Perspectives,* vol. 114, no. 9, September 2006, 1318–24; Rupa Basu and Jonathan M. Samut, "Relation Between Elevated Ambient Temperature and Mortality: A Review of the Epidemiologic Evidence," *Epidemiologic Reviews,* vol. 24, no. 2, 2002, 190–202; Anthony J. McMichael, Rosalie E. Woodruff, and Simon Hales, "Climate Change and Human Health: Present and Future Risks," *Lancet,* vol. 367, no. 9513, March 11, 2006, 858–69; W. J. M. Martens, "Climate Change, Thermal Stress, and Mortality Changes," *Social Science and Medicine,* vol. 46, no. 3, June 1998, 331–44; and W. R. Keatinge, G. C. Donaldson, Elvira Cordioli, et al., "Heat Related Mortality in Warm and Cold Regions of Europe: Observational Study," *British Medical Journal,* vol. 321, no. 7262, September 16, 2000, 670–73.

27. IPCC Fourth Assessment Report (2007): Working Group II: Impacts, Adaptation, and Vulnerability: Chapter 8, "Human Health," 391–431, last accessed on July 25, 2009, at http://www.ipcc.ch/pdf/assessment-report/ar4/wg2/ar4-wg2-chapter8.pdf.
28. Ibid., 396–413.
29. Ibid., 393.
30. Ibid.
31. Lomborg, *Cool It*, 15.
32. Ibid., 14.
33. Ibid., 170.
34. Ebi, Mills, Smith, and Grambsch, "Climate Change and Human Health Impacts in the United States: An Update on the Results of the U.S. National Assessment," *Environmental Health Perspectives*, vol. 114, no. 9, September 2006, 1318–24, last accessed on July 25, 2009, at http://www.ehponline.org/members/2006/8880/8880.pdf.
35. See U.S. Global Change Research Program: U.S. National Assessment of the Potential Consequences of Climate Variability and Change, last accessed on July 25, 2009, at http://www.usgcrp.gov/usgcrp/nacc/background/background.html.
36. "Suit Challenges Climate Change Report by U.S.," *New York Times*, August 7, 2003.
37. Ebi, Mills, Smith, and Grambsch, "Climate Change and Human Health Impacts in the United States: An Update on the Results of the U.S. National Assessment," 1318.
38. Basu and Samut, "Relation Between Elevated Ambient Temperature and Mortality: A Review of the Epidemiologic Evidence," *Epidemiological Reviews*, vol. 24, no. 2, 2002, 190, last accessed on July 25, 2009, at http://epirev.oxfordjournals.org/cgi/reprint/24/2/190.
39. McMichael, Woodruff, and Hales, "Climate Change and Human Health: Present and Future Risks," 859, 864.
40. Lomborg, *Cool It*, 14, 170.
41. Ibid.
42. Ibid., 14–15.
43. Keatinge, Donaldson, Cordioli, et al., "Heat Related Mortality in Warm and Cold Regions of Europe: Observational Study," 672, last accessed on July 25, 2009 at http://www.bmj.com/cgi/reprint/321/7262/670.
44. IPCC Fourth Assessment Report (2007): Working Group II: Impacts, Adaptation and Vulnerability: Chapter 8, "Human Health," 393.
45. Lomborg, *Cool It*, 13.

46. Ibid.
47. Janet Larsen, "Record Heat Wave in Europe Takes 35,000 Lives: Far Greater Losses May Lie Ahead," Earth Policy Institute, October 9, 2003, last accessed on July 25, 2009, at http://www.earth-policy.org/Updates/Update29.htm.
48. Lomborg, *Cool It*, 16.
49. Thomas N. Chase, Klaus Wolter, Roger A. Pielke Sr., and Ichtiaque Rasool, "Was the 2003 European Summer Heat Wave Unusual in a Global Context? *Geophysical Research Letters*, vol. 33, L23709, December 14, 2006.
50. Lomborg, *Cool It*, 16.
51. Jonathan A. Patz, Diarmid Campbell-Lendrum, Tracey Holloway, and Jonathan A. Foley, "Impact of Regional Climate Change on Human Health," *Nature*, vol. 438, November 17, 2005, 310.
52. On page 20 of *Cool It*, Lomborg wrote: "It is likely that for many cities the urban-heat-island increases of the twentieth century are on a larger scale than those to come from global warming in the twenty-first century. Yet the increases have not brought these cities tumbling down." On page 171, Lomborg cited "Patz et al., 2005:310." On page 231, Lomborg listed this source: "Patz, J. A., Campbell-Lendrum, D. Holloway, T., & Foley, J. A. (2005). Impact of Regional Climate Change on Human Health. *Nature*, 438(7066), 310–17."
53. IPCC Third Assessment Report (2001): Working Group II: Impacts, Adaptation and Vulnerability, "Technical Summary," 29, last accessed on July 25, 2009, at http://www.ipcc.ch/ipccreports/tar/wg2/pdf/wg2TARtechsum.pdf.
54. Ibid.
55. IPCC Fourth Assessment Report (2007): Working Group II: Impacts, Adaptation and Vulnerability: Chapter 12, "Europe," 544, last accessed on July 25, 2009, at http://www.ipcc.ch/pdf/assessment-report/ar4/wg2/ar4-wg2-chapter12.pdf.
56. Ibid., 548.
57. Ibid., 557.
58. Lomborg, *Cool It*, 17.
59. World Health Organization (WHO), *The World Health Report 2004: Changing History*, 2004, last accessed on July 25, 2009, at http://www.who.int/whr/2004/en/report04_en.pdf. Lomborg cited a chart on page 121 of the report; however, the extensive chart, "Annex Table 2," is located on pages 120–25, and includes no data on heat- or cold-related mortality.

60. Lomborg, *Cool It*, 170.
61. A search of the PDF file of WHO, *The World Health Report 2004: Changing History*, 2004, at http://www.who.int/whr/2004/en/report04_en.pdf, did not turn up any references to cold- and heat-related mortality.
62. Lomborg, *Cool It*, 170.
63. Ibid., 17.
64. Ibid.
65. W. R. Keatinge, G. C. Donaldson, Elvira Cordioli, et al., "Heat Related Mortality in Warm and Cold Regions of Europe: Observational Study," *British Medical Journal*, vol. 321, no. 7262, September 16, 2000, 670–73; W. R. Keatinge and G. C. Donaldson, "The Impact of Global Warming on Health and Mortality," *Southern Medical Journal*, vol. 97, no. 11, November 1, 2004, 1093–97; Ian H. Langford and Graham Bentham, "The Potential Effects of Climate Change on Winter Mortality in England and Wales," *International Journal of Biometeorology*, vol. 38, no. 3, September 1995, 141–47.
66. Lomborg, *Cool It*, 17.
67. W. J. M. Martens, "Climate Change, Thermal Stress and Mortality Changes," *Social Science and Medicine*, vol. 46, no. 3, February 1998, 331–44.
68. IPCC Fourth Assessment Report (2007): Working Group II: Impacts, Adaptation and Vulnerability: Chapter 8, "Human Health," 407.
69. Lomborg, *Cool It*, 38–39.
70. Francesco Bosello, Roberto Roson, and Richard S. J. Tol, "Economy-Wide Estimates of the Implications of Climate Change: Human Health," *Ecological Economics*, vol. 58, no. 3, June 25, 2006, 580.
71. IPCC Fourth Assessment Report (2007): Climate Change 2007: Synthesis Report, 49, last accessed on July 25, 2009, at http://www.ipcc.ch/pdf/assessment-report/ar4/syr/ar4_syr.pdf.
72. Ibid., 51.
73. Rupa Basu and Jonathan M. Samet, "Relation Between Elevated Ambient Temperature and Mortality: A Review of the Epidemiologic Evidence," *Epidemiologic Reviews*, vol. 24, no. 2, 2002, 190–202.
74. National Research Council, *Reconciling Observations of Global Temperature Change* (Washington, D.C.: National Academy Press, 2000), 86.
75. Basu and Samet, "Relation Between Elevated Ambient Temperature and Mortality," 190.
76. Anton E. Kunst, Casper W. N. Looman, and John P. Mackenbach, "Outdoor Air Temperature and Mortality in The Netherlands: A Time-Series

Analysis," *American Journal of Epidemiology*, vol. 137, no. 3, 1993, 331–41; Wen-Harn Pan, Lung-An Ti, and Ming-Jan Tsai, "Temperature Extremes and Mortality from Coronary Heart Disease and Cerebral Infarction in Elderly Chinese," *Lancet*, vol. 345, no. 8946, 1995, 353–55; C. H. Wyndham and S. A. Fellingham, "Climate and Disease," *South Africa Medical Journal*, vol. 53, no. 26, 1978, 1051–61; A. MacFarlane, "Daily Mortality and Environment in English Conurbations. II. Deaths During Summer Hot Spells in Greater London," *Environmental Research*, vol. 15, no. 3, June 1978, 332–41; Marc Saez, Jordi Sunyer, Jordi Castellsague, et al., "Relationship Between Weather Temperature and Mortality: A Times Series Analysis Approach in Barcelona," *International Journal of Epidemiology*, vol. 24, no. 3, June 1995, 576–82; and W. R. Keatinge, S. R. K. Coleshaw, and J. Holmes, "Changes in Seasonal Mortalities with Improvement in Home Heating in England and Wales from 1964 to 1984," *International Journal of Biometeorology*, vol. 33, no. 2, June 1989, 71–76.

77. Basu and Samet, "Relation Between Elevated Ambient Temperature and Mortality," 195.
78. Ibid.

Chapter 5. On Melting Glaciers and Rising Sea Levels

1. Bjørn Lomborg, *Cool It: The Skeptical Environmentalist's Guide to Global Warming* (New York: Knopf, 2007), 53.
2. IPCC Fourth Assessment Report (2007): Working Group I: The Physical Science Basis: Annex I: Glossary, 946, last accessed on July 26, 2009, at http://www.ipcc.ch/pdf/assessment-report/ar4/wg1/ar4-wg1-annexes.pdf.
3. Ibid., 944.
4. Ibid., 941.
5. IPCC Fourth Assessment Report (2007): Working Group I: The Physical Science Basis, "Technical Summary," 43, last accessed on July 26, 2009, at http://www.ipcc.ch/pdf/assessment-report/ar4/wg1/ar4-wg1-ts.pdf.
6. Ibid.
7. Ibid., 43–44.
8. IPCC Third Assessment Report (2001): Climate Change 2001: Synthesis Report: Summary For Policymakers, 6, last accessed on July 26, 2009, at http://www.ipcc.ch/pdf/climate-changes-2001/synthesis-spm/synthesis-spm-en.pdf.
9. IPCC Third Assessment Report (2001): Working Group I: The Scientific

Basis: Technical Summary, 30, last accessed on July 26, 2009, at http://www.ipcc.ch/ipccreports/tar/wg1/pdf/WG1_TAR-FRONT.PDF.

10. Ibid., 73–74.
11. IPCC Fourth Assessment Report (2007): Working Group I: The Physical Science Basis: Chapter 4, "Observations: Changes in Snow, Ice and Frozen Ground," 337–383, last accessed on July 26, 2009, at http://www.ipcc.ch/pdf/assessment-report/ar4/wg1/ar4-wg1-chapter4.pdf.
12. Ibid., 356.
13. IPCC Fourth Assessment Report (2007): Working Group I: The Physical Science Basis: Chapter 6, "Paleoclimate," 436, last accessed on July 26, 2009, at http://www.ipcc.ch/pdf/assessment-report/ar4/wg1/ar4-wg1-chapter6.pdf.
14. Ibid.
15. Lomborg, *Cool It*, 53–54.
16. Ibid., 53, 54, 55.
17. Ibid., 55.
18. IPCC Third Assessment Report (2001): Working Group I: The Scientific Basis: Chapter 2, "Observed Climate Variability and Change," 133–35, last accessed on July 26, 2009, at http://www.ipcc.ch/ipccreports/tar/wg1/pdf/TAR-02.PDF.
19. Ibid., 135.
20. Ibid.
21. IPCC Fourth Assessment Report (2007): Working Group I: The Physical Science Basis: Chapter 6, "Paleoclimate," 469.
22. Lomborg, *Cool It*, 55.
23. Ibid.
24. IPCC Fourth Assessment Report (2007): Working Group I: The Physical Science Basis: Chapter 6, "Paleoclimate," 461 (Box 6.3).
25. Ibid.
26. Lomborg, *Cool It*, 55.
27. J. Oerlemans, "Holocene Glacier Fluctuations: Is the Current Rate of Retreat Exceptional?" *Annals of Glaciology*, vol. 31, January 2000, 39 (Abstract), last accessed on July 26, 2009, at http://www.ingentaconnect.com/content/igsoc/agl/2000/00000031/00000001/art00008.
28. Lomborg, *Cool It*, 55–56.
29. Ibid., 56.
30. Georg Kaser, Douglas R. Hardy, and Thomas Mölg, et al., "Modern Glacier Retreat on Kilimanjaro as Evidence of Climate Change: Observations and Facts," *International Journal of Climatology*, vol. 24, no. 3, 2004, 329.

31. IPCC Fourth Assessment Report (2007): Working Group I: The Physical Science Basis: Chapter 4, "Observations: Changes in Snow, Ice and Frozen Ground," 360.
32. IPCC Fourth Assessment Report (2007): Working Group I: The Physical Science Basis, "Technical Summary," 43–46.
33. Lomborg, *Cool It*, 56.
34. Ibid., 57.
35. See, for example: IPCC Fourth Assessment Report (2007): Working Group II: Impacts, Adaptation and Vulnerability, "Technical Summary," 49, 59, 69, at http://www.ipcc.ch/pdf/assessment-report/ar4/wg2/ar4-wg2-ts.pdf; Working Group II: Chapter 1, "Assessment of Observed Changes and Responses in Natural and Managed Systems," 86, 88, at http://www.ipcc.ch/pdf/assessment-report/ar4/wg2/ar4-wg2-chapter1.pdf; Working Group II: Chapter 3, "Freshwater Resources and Their Management," 182–85, 187, at http://www.ipcc.ch/pdf/assessment-report/ar4/wg2/ar4-wg2-chapter3.pdf; Working Group II: Chapter 4, "Ecosystems: Their Properties, Goods and Services," 232–33, at http://www.ipcc.ch/pdf/assessment-report/ar4/wg2/ar4-wg2-chapter4.pdf; Working Group II: Chapter 8, "Human Health," 414, at http://www.ipcc.ch/pdf/assessment-report/ar4/wg2/ar4-wg2-chapter8.pdf; Working Group II: Chapter 10, "Asia," 483, 493, 494, at http://www.ipcc.ch/pdf/ assessment-report/ar4/wg2/ar4-wg2-chapter10.pdf. Each of these sources was last accessed on July 26, 2009.
36. Lomborg, *Cool It*, 58.
37. Frank Lehmkuhl and Lewis A. Owen, "Late Quaternary Glaciation of Tibet and the Bordering Mountains: A Review," *Boreas*, vol. 34, no. 2, May 2005, 87.
38. IPCC Fourth Assessment Report (2007): Working Group II: Impacts, Adaptation and Vulnerability: Chapter 10, "Asia," 493.
39. Christian Schneeberger, Heinz Blatter, Abe-Ouchi Ayako, and Martin Wild, "Modelling Changes in the Mass Balance of Glaciers of the Northern Hemisphere for a Transient 2 x CO2 Scenario," *Journal of Hydrology*, vol. 282, no. 1–4, 2003, 145–63.
40. Ibid., 148 (Table 1).
41. Lomborg, *Cool It*, 58.
42. Ibid., 58, 59.
43. IPCC Fourth Assessment Report (2007): Working Group II: Impacts, Adaptation and Vulnerability: Chapter 3, "Freshwater Resources and Their Management," 187.

44. IPCC Fourth Assessment Report (2007): Working Group II: Impacts, Adaptation and Vulnerability: Chapter 4, "Ecosystems: Their Properties, Goods and Services," 232.
45. IPCC Fourth Assessment Report (2007): Working Group II: Impacts, Adaptation and Vulnerability, "Technical Summary," 49.
46. IPCC Fourth Assessment Report (2007): Working Group II: Impacts, Adaptation and Vulnerability: Chapter 10, "Asia," 483.
47. IPCC Fourth Assessment Report (2007): Working Group II: Impacts, Adaptation and Vulnerability: Chapter 10, "Asia." See the projected impacts with respect to "Water Availability and Demand" (on Russia, Tajikistan, the Euphrates and Tigris Rivers, Lebanon, North China, the Mekong, and the aquaculture industry), "Water Quality," and "Implications of Droughts and Floods," 483–84.
48. Lomborg, *Cool It*, 59.
49. Ibid., 60.
50. "Flood Legend," *Encyclopedia Britannica*, http://www.britannica.com/ebi/article-9274347. The "Flood Legend" entry (9274347) in Encyclopedia Britannica online is available with a free trial subscription.
51. Lomborg, *Cool It*, 60.
52. Bill McKibben, "The Submerging World," *Orion*, September/October 2004, last accessed on July 26, 2009, at http://www.orionmagazine.org/index.php/articles/article/146/.
53. Ibid.
54. Lomborg, *Cool It*, 60.
55. Ibid.
56. IPCC Fourth Assessment Report (2007): Working Group I: The Physical Science Basis, "Technical Summary," 70 (Table TS.6).
57. IPCC Fourth Assessment Report (2007): Working Group I: The Physical Science Basis: Climate Change 2007: Synthesis Report, 45, last accessed on July 26, 2009, at http://www.ipcc.ch/pdf/assessment-report/ar4/syr/ar4_syr.pdf.
58. Ibid., 45, 73.
59. IPCC Fourth Assessment Report (2007): Working Group I: The Scientific Basis: Annex I: Glossary, 946.
60. H. Jay Zwally, Waleed Abdalati, Tom Herring, et al., "Surface Melt-Induced Acceleration of Greenland Ice-Sheet Flow," *Science*, vol. 297, no. 5579, July 12, 2002, 218.
61. Ibid., 221.
62. "Melting Greenland Glacier May Hasten Rise in Sea Level," *Independent*, July 25, 2005.

63. Ibid.
64. "Sea Levels Likely to Rise Much Faster Than Was Predicted," *Independent*, February 17, 2006. See also, "Turning Point Warning as Greenland's Ice Disappears," *Guardian*, February 17, 2006.
65. "Climate Change: On The Edge: Greenland Ice Cap Breaking Up at Twice the Rate It Was Five Years Ago, Says Scientist Bush Tried to Gag," *Independent*, February 17, 2006.
66. Ibid.
67. "Greenland's Ice Cap Is Melting at a Frighteningly Fast Rate," *San Francisco Chronicle*, August 11, 2006.
68. Ibid.
69. Lomborg, *Cool It*, 60.
70. IPCC Fourth Assessment Report (2007): Working Group I: The Physical Science Basis: Chapter 5, "Observations: Oceanic Climate Change and Sea Level," 409, last accessed on July 26, 2009, at http://www.ipcc.ch/pdf/assessment-report/ar4/wg1/ar4-wg1-chapter5.pdf.
71. "The Weather Turns Wild: Global Warming Could Cause Droughts, Disease, and Political Upheaval," *U.S. News and World Report*, January 28, 2001.
72. Lomborg, *Cool It*, 61.
73. Ibid., 61, 179, 226. The travel article referenced by Lomborg is Neal Matthews, "Florida: The Attack of the Killer Architects," *Travel Holiday*, September 1, 2000, last accessed on July 26, 2009, at http://nealmatthews.com/Documents/Miami%20architecture.mht.
74. IPCC Third Assessment Report (2001): Working Group II: Impacts, Adaptation and Vulnerability: Chapter 15, "North America," 766, last accessed on July 26, 2009, at http://www.ipcc.ch/ipccreports/tar/wg2/pdf/wg2TARchap15.pdf.
75. Lomborg, *Cool It*, 62.
76. Gore, *An Inconvenient Truth*, 190–97.
77. Ibid., 190.
78. Ibid., 192–95.
79. Ibid., 190–209.
80. Ibid., 196–97.
81. "Climate Change: On The Edge," *Independent*, February 17, 2006.
82. Ibid.

Chapter 6. On Greenland and the Missing Figures

1. Bjørn Lomborg, *Cool It: The Skeptical Environmentalist's Guide to Global Warming* (New York: Knopf, 2007), 62.
2. IPCC Fourth Assessment Report (2007): Working Group I: The Physical Science Basis: Chapter 10, "Global Climate Projections," 812 (Table 10.31), last accessed on July 27, 2009, at http://www.ipcc.ch/pdf/assessment-report/ar4/wg1/ar4-wg1-chapter10.pdf.
3. Lomborg, *Cool It*, 62–63.
4. IPCC Fourth Assessment Report (2007): Working Group I: The Physical Science Basis: Chapter 10, "Global Climate Projections," 814.
5. Ibid., 814–16.
6. Lomborg, *Cool It*, 63.
7. IPCC Fourth Assessment Report (2007): Working Group I: The Physical Science Basis: Chapter 10, "Global Climate Projections," 816.
8. Ibid., 816–20.
9. Lomborg, *Cool It*, 63.
10. Ibid.
11. Ibid., 63–64.
12. J. Oerlemans, R. P. Bassford, W. Chapman, et al., "Estimating the Contribution of Arctic Glaciers to Sea-Level Change in the Next 100 Years," *Annals of Glaciology*, vol. 42, no. 1, 2005, 230–36.
13. J. E. Hansen, "Scientific Reticence and Sea Level Rise," *Environmental Research Letters*, May 24, 2007, last accessed on July 27, 2009, at http://www.iop.org/EJ/article/1748-9326/2/2/024002/erl7_2_024002.html.
14. Isabella Velicogna and John Wahr, "Acceleration of Greenland Ice Mass Loss in Spring 2004," *Nature*, vol. 443, September 21, 2006, 329.
15. J. L. Chen, C. R. Wilson, and B. D. Tapley, "Satellite Gravity Measurements Confirm Accelerated Melting of Greenland Ice Sheet," *Science*, vol. 313, no. 5795, September 29, 2006, 1958.
16. S. B. Luthcke, H. J. Zwally, W. Abdalati, et al., "Recent Greenland Ice Mass Loss by Drainage System from Satellite Gravity Observations," *Science*, vol. 314, no. 5803, November 24, 2006, 1289.
17. Andrew Shepherd and Duncan Wingham, "Recent Sea-Level Contributions of the Antarctic and Greenland Ice Sheets," *Science*, vol. 315, no. 5818, March 16, 2007, 1529.
18. J. Oerlemans, R. P. Bassford, W. Chapman, et al., "Estimating the Contribution of Arctic Glaciers to Sea-Level Change in the Next 100 Years," *Annals of Glaciology*, vol. 42, no. 1, 2005, 230.

19. Lomborg, *Cool It*, 64.
20. J. M. Gregory and P. Huybrechts, "Ice-Sheet Contributions to Future Sea-Level Change," *Philosophical Transactions of the Royal Society A*, vol. 364, no. 1844, 2006, 1710.
21. Lomborg, *Cool It*, 64.
22. IPCC Fourth Assessment Report (2007): Working Group I: The Physical Science Basis: Chapter 10, "Global Climate Projections," 818.
23. H. Jay Zwally, Waleed Abdalati, Tom Herring, et al., "Surface Melt-Induced Acceleration of Greenland Ice-Sheet Flow," *Science*, vol. 297, no. 5579, July 12, 2002, 218–22.
24. H. Jay Zwally, Mario B. Giovinetto, Jun Li, et al., "Mass Changes of the Greenland and Antarctic Ice Sheets and Shelves and Contributions to Sea-Level Rise: 1992–2002," *Journal of Glaciology*, vol. 51, no. 175, December 2005, 509–27.
25. Velicogna and Wahr, "Acceleration of Greenland Ice Mass Loss in Spring 2004." 329–31.
26. Chen, Wilson, and Tapley, "Satellite Gravity Measurements Confirm Accelerated Melting of Greenland Ice Sheet," 1958–60.
27. Luthcke, Zwally, Abdalati, et al., "Recent Greenland Ice Mass Loss by Drainage System from Satellite Gravity Observations." 1286–89.
28. See IPCC Fourth Assessment Report (2007): Working Group I: The Physical Science Basis: Chapter 10, "Global Climate Projections," 817.
29. Ibid.
30. Hansen, "Scientific Reticence and Sea Level Rise."
31. Ibid.
32. IPCC Fourth Assessment Report (2007): Working Group I: The Physical Science Basis: Annex I: Glossary, 950, last accessed on July 27, 2009, at http://www.ipcc.ch/pdf/assessment-report/ar4/wg1/ar4-wg1-annexes.pdf.
33. Hansen, "Scientific Reticence and Sea Level Rise."
34. IPCC Fourth Assessment Report (2007): Working Group I: The Physical Science Basis: Chapter 10 "Global Climate Projections," 749.
35. Hansen, "Scientific Reticence and Sea Level Rise."

Chapter 7. The Penguins Sidebar

1. Bjørn Lomborg, *Cool It: The Skeptical Environmentalist's Guide to Global Warming* (New York: Knopf, 2007), 66.
2. Ibid.

3. Christophe Barbraud and Henri Weimerskirch, "Emperor Penguins and Climate Change," *Nature*, vol. 411, May 10, 2001, 183.
4. Lomborg, *Cool It*, 66.
5. Australia Antarctic Government Division, "Where Do They Breed?" This document was originally accessed at http://www.aad.gov.au/default.asp?casid=2879. As of July 27, 2009, it does not appear to be available on the Internet. A hard copy is available upon request.
6. Lomborg, *Cool It*, 66.
7. E. J. Woehler and J. P. Croxall, "The Status and Trends of Antarctic and Sub-Antarctic Seabirds," *Marine Ornithology*, vol. 25, no. 1 and 2, 1997, 43–66.
8. Ibid., 44.
9. Ibid.
10. Ibid., 45 (Table 1).
11. Lomborg, *Cool It*, 66.
12. The IUCN Red List of Threatened Species: *Aptenodytes forsteri*, 2008, last accessed on July 27, 2009, at http://www.iucnredlist.org/details/144799.
13. IPCC Fourth Assessment Report (2007): Working Group II: Impacts, Adaptation and Vulnerability: Chapter 15, "Polar Regions (Arctic and Antarctic)," 660, last accessed on July 27, 2009, at http://www.ipcc.ch/pdf/assessment-report/ar4/wg2/ar4-wg2-chapter15.pdf.
14. Barbraud and Weimerskirch, "Emperor Penguins and Climate Change," 183.
15. Lomborg, *Cool It*, 66.
16. Stephanie Jenouvrier, Christophe Barbraud, and Henri Weimerskirch, "Sea Ice Affects the Population Dynamics of Adélie Penguins in Terre Adélie," *Polar Biology*, vol. 29, no. 5, 2006, 413.
17. See the undated video at "Penguin Science: Understanding Penguin Response to Climate and Ecosystem Change," last accessed on July 27, 2009, at http://www.penguinscience.com/media/video/webisodes.php.
18. "Antarctic Penguins and Climate Change: Real Leaders Tackle Climate Change," World Wildlife Fund, December 2007, last accessed on July 27, 2009, at http://assets.panda.org/downloads/folleto_penguins.pdf.
19. Ibid.
20. British Antarctic Survey: Natural Environment Research Council, "Science Briefing—Penguins," 2008, last accessed on July 27, 2009, at http://

www.antarctica.ac.uk/press/journalists/resources/science/sciencebriefi ngpenguins.php.

21. Celine Le Bohec, Joël M. Durant, Michel Gauthier-Clerc, et al., "King Penguin Population Threatened by Southern Ocean Warming," *Proceedings of the National Academy of Sciences*, vol. 105, no. 7, February 19, 2008, 2493.

22. See, for example: Barbraud and Weimerskirch, "Emperor Penguins and Climate Change," *Nature*, vol. 411, May 10, 2001, 183–86; Christophe Barbraud and Henri Weimerskirch, "Antarctic Birds Breed Later in Response to Climate Change," *Proceedings of the National Academy of Sciences*, vol. 103, no. 16, April 18, 2006, 6248–51; Le Bohec, Durant, Gauthier-Clerc, et al., "King Penguin Population Threatened by Southern Ocean Warming"; "Arctic Penguins and Climate Change," World Wildlife Fund, December 2007; "Half the World's Penguin Species Marching Towards Extinction: Conservation Group Seeks Protection from Effects of Global Warming," Center for Biological Diversity, November 28, 2006; "Agency Takes First Step to Protect Emperor Penguins and 9 Others," *New York Times*, July 11, 2007; "Conservation Group Sues Bush Administration Over Delay in Penguin Protection: 10 Species Including Emperor Penguin Threatened by Global Warming," Center for Biological Diversity, February 27, 2008. See also the IPCC Fourth Assessment Report (2007): Working Group II: Impacts, Adaptation and Vulnerability, "Technical Summary," 45, at http://www.ipcc.ch/pdf/assessment-report/ar4/wg2/ar4-wg2-ts.pdf: "As sea-ice biomes shrink, dependent polar species, including predators such as penguins, seals and polar bears, are very likely to experience habitat degradation and losses."

Chapter 8. On Antarctica and the Larsen-B Ice Shelf

1. Bjørn Lomborg, *Cool It: The Skeptical Environmentalist's Guide to Global Warming* (New York: Knopf, 2007), 64–67.

2. William L. Chapman and John E. Walsh, "A Synthesis of Antarctic Temperatures," Department of Atmospheric Sciences, University of Illinois at Urbana-Champaign (unpublished draft, March 2005), last accessed on July 28, 2009, at http://igloo.atmos.uiuc.edu/Antarctic.paper.chapwalsh .2005.pdf; O. Humlum, "Antarctic Temperature Changes During the Observational Period, a Project Funded by the University Courses on Svalbard (UNIS), 2001–2005," Department of Geology, Svalbard, Norway, April 2005, last accessed on July 28, 2009, at http://www.unis.no/

studies/geology/ag_204_more_info/ole/AntarcticTemperatureChanges.htm; Andrew J. Monaghan and David H. Bromwich, "A High Spatial Resolution Record of Near-surface Temperature over WAIS [Western Antarctic Ice Sheet] During the Past 5 Decades," Thirteenth Annual WAIS Workshop, 2006.

3. Chapman and Walsh, "A Synthesis of Antarctic Temperatures," *Journal of Climate (American Meteorological Society)*, vol. 20, no. 16, August 2007, 4,096.
4. Chapman and Walsh, "A Synthesis of Antarctic Temperatures," unpublished draft, 2.
5. Ibid.
6. Humlum, "Antarctic Temperature Changes During the Observational Period: A Project Funded by the University Courses on Svalbard (UNIS) 2001–2005."
7. See Lomborg, *Cool It*, 65, 181, and 227, for references to Monaghan and Bromwich, "A High Spatial Resolution Record of Near-surface Temperature over WAIS [Western Antarctica Ice Sheet] during the Past 5 Decades."
8. Andrew J. Monaghan and David H. Bromwich, "Advances in Describing Recent Antarctic Climate Variability," *Bulletin of the American Meteorological Society*, vol. 89, no. 9, September 2008, 1, 297.
9. Andrew J. Monaghan, David H. Bromwich, William Chapman, and Josefino C. Comiso, "Recent Variability and Trends of Antarctic Near-Surface Temperature," *Journal of Geophysical Research*, vol. 113, February 22, 2008, Abstract, last accessed on July 28, 2009, at http://www.agu.org/pubs/crossref/2008/2007JD009094.shtml.
10. Nathan P. Gillett, Dáithí A. Stone, Peter A. Stott, et al., "Attribution of Polar Warming to Human Influence," *Nature Geoscience*, published online October 30, 2008, Abstract, last accessed on July 28, 2009, at http://www.nature.com/ngeo/journal/v1/n11/abs/ngeo338.html.
11. David P. Schneider and Eric J. Steig, "Ice Cores Record Significant 1940s Antarctic Warmth Related to Tropical Climate Variability," *Proceedings of the National Academy of Sciences*, vol. 105, no. 34, August 26, 2008, 12,154.
12. Eric J. Steig, David P. Schneider, Scott D. Rutherford, et al., "Warming of the Antarctic Ice-Sheet Surface Since the 1957 International Geophysical Year," *Nature*, vol. 457, January 22, 2009, Abstract, last accessed on July 28, 2009, at http://www.nature.com/nature/journal/v457/n7228/full/nature07669.html.
13. Lomborg, *Cool It*, 65.
14. Ibid.
15. Ibid., 181.

16. Al Gore, *An Inconvenient Truth: The Planetary Emergency of Global Warming and What We Can Do About It* (New York: Rodale, 2006), 182–83.
17. Ibid., 184–85.
18. For detailed discussions concerning the overall accuracy of *An Inconvenient Truth*, see Eric Steig, "Al Gore's Movie," realclimate.org, May 10, 2006, last accessed on July 28, 2009, at http://www.realclimate.org/index.php/archives/2006/05/al-gores-movie/; "Al Gore's An Inconvenient Truth: Unscientific?" *New Scientist*, October 12, 2007; "U.K. Judge Rules Gore's Climate Film Has 9 Errors," *Washington Post*, October 12, 2007; Gavin Schmidt and Michael Mann, "Convenient Truths," realclimate.org, October 15, 2007, last accessed on July 28, 2009, at http://www.realclimate.org/index.php/archives/2007/10/convenient-untruths/.
19. C. S. M. Doake, H. F. J. Corr, H. Rott, et al., "Breakup and Conditions for Stability of the Northern Larsen Ice Shelf, Antarctica," *Nature*, vol. 391, February 19, 1998, 778, Abstract, last accessed on July 28, 2009, at http://www.nature.com/nature/journal/v391/n6669/abs/391778a0.html.
20. Ted Scambos, "Breakup of Larsen-B Ice Shelf May Be Underway," National Snow and Ice Data Center, 1998, last accessed on July 29, 2009, at http://nsidc.org/news/press/larsen_B/1998.html.
21. David G. Vaughan, Gareth J. Marshall, William M. Connolley, et al., "Devil in the Detail," *Science*, vol. 293, no. 5536, September 7, 2001, 1778.
22. "Antarctic Ice Shelf Breaks Apart," BBC News, March 19, 2002.
23. Gore, *An Inconvenient Truth*, 184.
24. Hernán De Angelis and Pedro Skvarca, "Glacier Surge After the Shelf Collapse," *Science*, vol. 299, no. 5612, March 7, 2003, 1560, 1561.
25. Tom Clarke, "Sea-Level Rise Shelved For Now: Glaciers Spotted Surging to Sea After Ice-Shelf Collapse," *Nature News*, March 7, 2003.
26. "Antarctic Ice Shelf Disintegration Underscores a Warming World," National Snow and Ice Data Center, March 25, 2008, last accessed on July 28, 2009, at http://nsidc.org/news/press/20080325_Wilkins.html; "Disintegration: Antarctic Warming Claims Another Ice Shelf," NASA Earth Observatory, March 26, 2008, last accessed on July 28, 2009, at http://earthobservatory.nasa.gov/Features/WilkinsIceSheet/.
27. David P. Schneider and Eric J. Steig, "Ice Cores Record Significant 1940s Antarctic Warmth Related to Tropical Climate Variability."
28. T. A. Scambos, J. A. Bohlander, C. A. Shuman, and P. Skvarca, "Glacier Acceleration and Thinning After Ice Shelf Collapse in the Larsen-B Em-

bayment, Antarctica," *Geophysical Research Letters*, vol. 31, no. 18, L18402, September 22, 2004, Abstract, last accessed on July 28, 2009, at http://www.agu.org/pubs/crossref/2004/2004GL020670.shtml.

29. Gore, *An Inconvenient Truth*, 184.
30. National Snow and Ice Data Center, "Antarctic Glaciers Speed Up," December 9, 2003, last accessed on July 28, 2009, at http://nsidc.org/news/press/20031219_speed_up.html.
31. Laura Naranjo, "After the Larsen-B," NASA Earth System Science Data and Services, December 5, 2007, last accessed on July 28, 2009, at http://nasadaacs.eos.nasa.gov/articles/2007/2007_larsen.html.
32. Lomborg, *Cool It*, 65.
33. Ibid.
34. Ibid., 181.
35. Ibid., 65.
36. Carol J. Pudsey, John W. Murray, Peter Appleby, and Jeffrey Evans, "Ice Shelf History From Petrographic and Foraminiferal Evidence, Northeast Antarctic Peninsula," *Quaternary Science Reviews*, vol. 25, no. 17–18, September 2006, 2357–79.
37. Ibid., 2357.
38. Lomborg, *Cool It*, 65.
39. Ibid., 65, 181. Lomborg's reference note reads: "Pudsey, et al., 2006:2375; Vaughan et al., 2001."
40. Vaughan, Marshall, Connolley, et al., "Devil in the Detail," 1777–79.
41. Ibid., 1778.
42. Ibid.
43. Eugene Domack, Diana Duran, Amy Leventer, et al., "Stability of the Larsen-B Ice Shelf on the Antarctic Peninsula During the Holocene Epoch," *Nature*, vol. 436, August 4, 2005, 681.
44. Ibid.
45. N. F. Glasser and T. A. Scambos, "A Structural Glaciological Analysis of the 2002 Larsen-B Ice Shelf Collapse," *Journal of Glaciology*, vol. 54, no. 184, 2008, 2–16.
46. "Antarctic Ice Shelf Collapse Blamed on More Than Global Warming," *ScienceDaily*, February 11, 2008.
47. Ibid.
48. Gore, *An Inconvenient Truth*, 182–83.
49. Lomborg, *Cool It*, p. 65.
50. Pudsey, Murray, Appleby, and Evans, "Ice Shelf History from Petro-

graphic and Foraminiferal Evidence, Northeast Antarctic Peninsula," 2357–79.

51. Vaughan, Marshall, Connolley, et al., "Devil in the Detail," 1777–79.
52. Pudsey, Murray, Appleby, and Evans, "Ice Shelf History From Petrographic and Foraminiferal Evidence, Northeast Antarctic Peninsula," 2357.
53. Lomborg, *Cool It*, 65.
54. Gore, *An Inconvenient Truth*, 184.
55. Lomborg, *Cool It*, 65–66.
56. Ibid., 66.
57. Lomborg cited four studies to support this sentence: John Turner, Tom Lachlan-Cope, Steve Colwell, and Gareth J. Marshall, "A Positive Trend in Western Antarctic Peninsula Precipitation over the Last 50 Years Reflecting Regional and Antarctic-wide Atmospheric Circulation Changes," *Annals of Glaciology*, vol. 41, no. 1, June 2005, 85–91; D. J. Wingham, A. Shepherd, A. Muir, and G. J. Marshall, "Mass Balance of the Antarctic Ice Sheet," *Philosophical Transactions of the Royal Society A*, vol. 364, 2006, 1627–35; H. Jay Zwally, Mario B. Giovinetto, Jun Li, et al., "Mass Changes of the Greenland and Antarctic Ice Sheets and Contributions to Sea-Level Rise: 1992–2002," *Journal of Glaciology*, vol. 51, no. 175, December 2005, 509–27; and Elizabeth M. Morris and Robert Mulvaney, "Recent Variations in Surface Mass Balance of the Antarctic Peninsula Ice Sheet," *Journal of Glaciology*, vol. 50, no. 169, March 2004, 257–67.
58. H. D. Pritchard and D. G. Vaughan, "Widespread Acceleration of Tidewater Glaciers on the Antarctic Peninsula," *Journal of Geophysical Research*, vol. 112, June 6, 2007, Abstract, last accessed on July 28, 2009, at http://www.agu.org/pubs/crossref/2007/2006JF000597.shtml.

Chapter 9. On Hurricanes and Extreme Weather Events

1. Bjørn Lomborg, *Cool It: The Skeptical Environmentalist's Guide to Global Warming* (New York: Knopf, 2007), 72.
2. Ibid., 73.
3. Al Gore, *An Inconvenient Truth: The Planetary Emergency of Global Warming and What We Can Do About It* (New York: Rodale, 2006), 96–99.
4. Ibid., 92–99.
5. Lomborg, *Cool It*, 73.
6. Kerry Emanuel, "Increasing Destructiveness of Tropical Cyclones Over the Past 30 Years," *Nature*, vol. 436, August 4, 2005, 686.

7. Robert F. Kennedy Jr., "'For They That Sow the Wind Shall Reap the Whirlwind,'" Huffington Post, August 29, 2005, last accessed on July 28, 2009, at http://www.huffingtonpost.com/robert-f-kennedy-jr/for-they-that-sow-the-_b_6396.html.
8. Gore, *An Inconvenient Truth*, 92.
9. With respect to global warming and hurricane intensity, see also: Kerry Emanuel, Ragoth Sundararajan, and John Williams, "Hurricanes and Global Warming: Results from Downscaling IPCC AR4 Simulations," *Bulletin of the American Meteorological Society*, vol. 89, no. 3, March 2008, 347–67; Kerry Emanuel, "The Hurricane-Climate Connection," *Bulletin of the American Meteorological Society*, vol. 89, no. 5, May 2008, ES10–ES20.
10. IPCC Third Assessment Report (2001): Working Group II: Impacts, Adaptation and Vulnerability, "Technical Summary," 50, last accessed on July 28, 2009, at http://www.ipcc.ch/ipccreports/tar/wg2/pdf/wg2TARtechsum.pdf.
11. Ibid., 72.
12. Ibid.
13. IPCC Third Assessment Report (2001): Climate Change 2001: Synthesis Report: Summary for Policymakers, 14, last accessed on July 28, 2009, at http://www.ipcc.ch/ipccreports/tar/vol4/pdf/spm.pdf.
14. IPCC Fourth Assessment Report (2007): Working Group I: The Physical Science Basis, "Technical Summary," Box TS.5: "Extreme Weather Events," 53, last accessed on July 28, 2009, at http://www.ipcc.ch/pdf/assessment-report/ar4/wg1/ar4-wg1-ts.pdf.
15. IPCC Fourth Assessment Report (2007): Working Group I: The Physical Science Basis: Chapter 3, "Observations: Surface and Atmospheric Climate Change," "Frequently Asked Question 3.3: Has There Been a Change in Extreme Events like Heat Waves, Droughts, Floods and Hurricanes?" 308, last accessed on July 28, 2009, at http://www.ipcc.ch/pdf/assessment-report/ar4/wg1/ar4-wg1-chapter3.pdf.
16. Ibid., 310–12.
17. Ibid., 312.
18. Gore, *An Inconvenient Truth*, 88–89.
19. Lomborg, *Cool It*, 73.
20. Ibid., 81.
21. Ibid.
22. Ibid.
23. Ibid.

24. Pavel Ya. Groisman, Richard W. Knight, David R. Easterling, et al., "Trends in Intense Precipitation in the Climate Record," *Journal of Climate*, vol. 18, no. 9, May 2005, 1344, 1326, last accessed on July 28, 2009, at http://www.meteo.mcgill.ca/wise/Groisman_et_al_2005_Intense_precip.pdf.
25. IPCC Fourth Assessment Report (2007): Working Group I: The Physical Science Basis: Chapter 10, "Global Climate Projections," 783, last accessed on July 28, 2009, at http://www.ipcc.ch/pdf/assessment-report/ar4/wg1/ar4-wg1-chapter10.pdf.
26. Lomborg, *Cool It*, 81.
27. P. C. D. Milly, R. T. Wetherald, K. A. Dunne, and T. L. Delworth, "Increasing Risk of Great Floods in a Changing Climate," *Nature*, vol. 415, January 31, 2002, 514 (Abstract), last accessed on July 29, 2009, at http://www.nature.com/nature/journal/v415/n6871/full/415514a.html.
28. Lomborg, *Cool It*, 184, 227.
29. Ibid., 81.
30. IPCC Fourth Assessment Report (2007): Working Group I: The Physical Science Basis, "Frequently Asked Question 9.1: Can Individual Extreme Events be Explained by Greenhouse Warming?" 119, last accessed on July 28, 2009, at http://www.ipcc.ch/pdf/assessment-report/ar4/wg1/ar4-wg1-faqs.pdf.
31. Ibid.
32. Lomborg, *Cool It*, 81–82.
33. "Dozens Are Dead As Floods Sweep Through Europe," *New York Times*, August 13, 2002; "Tens of Thousands Flee Prague as Floods Invade Historic Center," *New York Times*, August 14, 2002; "As Floods Ebb in Prague, Threat Rolls into Germany," *New York Times*, August 15, 2002.
34. IPCC Fourth Assessment Report (2007): Working Group II: Impacts, Adaptation and Vulnerability, "Technical Summary," 35, last accessed on July 28, 2009, at http://www.ipcc.ch/pdf/assessment-report/ar4/wg2/ar4-wg2-ts.pdf.
35. Ibid., 44 (Box TS.5).
36. Ibid., 49.
37. Ibid., 53.
38. Ibid., 59 (Box TS.6).
39. Lomborg, *Cool It*, 82.
40. Zbigniew W. Kundzewicz, Dariusz Graczyk, Thomas Maurer, et al., "Trend Detection in River Flow Series: 1. Annual Maximum Flow," *Hy-*

drological Sciences Journal, vol. 50, no. 5, October 2005, 797–810; Cecilia Svensson, Zbigniew W. Kundzewicz, and Thomas Maurer, "Trend Detection in River Flow Series: 2. Flood and Low-Flow Index Series," *Hydrological Sciences Journal*, vol. 50, no. 5, October 2005, 811–24.

41. Lomborg, *Cool It*, 82.
42. IPCC Fourth Assessment Report (2007): Working Group II: Impacts, Adaptation, and Vulnerability: Chapter 10, "Asia," 476 (Table 10.3), last accessed on July 28, 2009, at http://www.ipcc.ch/pdf/assessment-report/ar4/wg2/ar4-wg2-chapter10.pdf.
43. Lomborg, *Cool It*, 81–87.
44. IPCC Fourth Assessment Report (2007): Working Group II: Impacts, Adaptation, and Vulnerability: Chapter 13, "Latin America," 583, last accessed on July 28, 2009, at http://www.ipcc.ch/pdf/assessment-report/ar4/wg2/ar4-wg2-chapter13.pdf.
45. Ibid., 585.

Chapter 10. Malaria in Vermont

1. A. J. McMichael, D. H. Campbell-Lendrum, C. F. Corvalán, et al. (eds.), *Climate Change and Human Health: Risks and Responses: Summary*, World Health Organization, 2003, last accessed on July 29, 2009, at http://www.who.int/globalchange/publications/climchange.pdf.
2. Bjørn Lomborg, *Cool It: The Skeptical Environmentalist's Guide to Global Warming* (New York: Knopf, 2007), 92.
3. McMichael, Campbell-Lendrum, Corvalan, et al. (eds.), *Climate Change and Human Health*, Chapter 13, "Conclusions and Recommendations for Action," 276.
4. Lomborg, *Cool It*, 92–93.
5. Ibid., 187.
6. "Bangladesh Is Paying a Cruel Price for the West's Excesses," *Guardian*, December 7, 2006.
7. Lomborg listed the source as follows: "Consultation Paper on Climate Change. UK Liberal Democrats," 2006. Retrieved 1-1-07, from http://consult.libdems.org.uk/climatechange/wp-content/uploads/2006/09/climate-change-cp84.pdf." See Lomborg, *Cool It*, 92–93, 187, 224.
8. "Two-Thirds of Energy Wasted by Antiquated UK System," Greenpeace, U.K., July 19, 2005, last accessed on July 29, 2009, at http://www.greenpeace.org.uk/media/press-releases/two-thirds-of-energy-wasted-by-antiquated-uk-system.

9. Lomborg, *Cool It*, 93.
10. "Climate Change Death Toll Put at 150,000," Reuters (via Common Dreams), December 11, 2003.
11. See the Common Dreams News Center at http://www.commondreams.org.
12. Lomborg, *Cool It*, 93.
13. McMichael, Campbell-Lendrum, Corvalan, et al. (eds.), *Climate Change and Human Health*, Chapter 7, "How Much Disease Could Climate Change Cause?" 139–40.
14. Lomborg, *Cool It*, 93.
15. McMichael, Campbell-Lendrum, Corvalan, et al. (eds.), *Climate Change and Human Health*, Chapter 7, "How Much Disease Could Climate Change Cause?" 153.
16. Ibid., 133–55.
17. Lomborg, *Cool It*, 93.
18. Ibid., 187.
19. Lomborg's citation to this table reads in its entirety: "CRU, (2006). HadCRUT3 Temperature: Global. Climatic Research Unit, University of East Anglia. Retrieved 1-1-07, from http://www.cru.uea.ac.uk/cru/data/temperature/crutem3gl.txt." I last accessed this URL, which can be located on page 210 in *Cool It*, on July 29, 2009.
20. "Climate Change and Human Health: Risks and Responses: Summary," WHO, WMO, and UNEP, 2003, 7, last accessed on July 29, 2009, at http://www.who.int/globalchange/climate/en/ccSCREEN.pdf.
21. Ibid.
22. Lomborg, *Cool It*, 94.
23. Kofi Annan, "Citing 'Frightening Lack of Leadership' on Climate Change, Secretary-General Calls Phenomenon an All-Encompassing Threat in Address to Nairobi Talks," United Nations, Department of Public Information, News and Media Division, New York, November 15, 2006, last accessed on July 29, 2009, at http://www.un.org/News/Press/docs/2006/sgsm10739.doc.htm.
24. Ibid.
25. Lomborg, *Cool It*, 94.
26. IPCC Third Assessment Report (2001): Working Group II: Impacts, Adaptation, and Vulnerability, "Technical Summary," 59, last accessed on July 29, 2009, at http://www.ipcc.ch/ipccreports/tar/wg2/pdf/wg2TARtechsum.pdf.
27. IPCC Third Assessment Report (2001): Working Group II: Impacts,

Adaptation, and Vulnerability: Chapter 9, "Human Health," 465, last accessed on July 29, 2009, at http://www.ipcc.ch/ipccreports/tar/wg2/pdf/wg2TARchap9.pdf.

28. "The Weather Turns Wild: Global Warming Could Cause Droughts, Disease, and Political Upheaval," *U.S. News and World Report*, January 28, 2001.
29. Lomborg, *Cool It*, 94.
30. P. Martens, R. S. Kovats, S. Nijhof, et al., "Climate Change and Future Populations at Risk of Malaria," *Global Environmental Change*, vol. 9, supp. 1, October 1999, S89.
31. M. Van Lieshout, R. S. Kovats, M. T. J. Livermore, and P. Martens, "Climate Change and Malaria: Analysis of the SRES Climate and Socio-Economic Scenarios," *Global Environmental Change*, vol. 14, no. 1, April 2004, 92.
32. IPCC Fourth Assessment Report (2007): Working Group II: Impacts, Adaptation, and Vulnerability, "Technical Summary," 47, last accessed On July 29, 2009, at http://www.ipcc.ch/pdf/assessment-report/ar4/wg2/ar4-wg2-ts.pdf.
33. Lomborg, *Cool It*, 188.
34. David A. King, "Policy Forum: Environment: Climate Change Science: Adapt, Mitigate, or Ignore?" *Science*, vol. 303, no. 5655, January 9, 2004, 176–77.
35. Lomborg, *Cool It*, 43, 100, 102, 164.
36. Ibid., 164.

Chapter 11. On Malnutrition

1. Bjørn Lomborg, *Cool It: The Skeptical Environmentalist's Guide to Global Warming* (New York: Knopf, 2007), 102–3.
2. "Global Warming: Scientists Reveal Timetable," the *Independent* (via Common Dreams, as referenced by Lomborg), February 3, 2005.
3. Ibid.
4. Lomborg, *Cool It*, 103.
5. "Global Warming Will Increase World Hunger," Reuters (via Global Policy Forum, as referenced by Lomborg), May 27, 2005.
6. Ibid.
7. Lomborg, *Cool It*, 103.
8. Ibid., 60–67.
9. Ibid., 103.

10. Ibid.
11. The seven sources cited are: (a) "World Agriculture: Towards 2030/2050: Interim Report: Prospects for Food, Nutrition, Agriculture and Major Commodity Groups: Global Perspectives Studies Unit: Food and Agriculture Organization of the United Nations: Rome, June 2006," 8, at http://www.fao.org/es/ESD/AT2050web.pdf; (b) "FAOSTAT: Food and Agriculture Organization of the United Nations," at http://faostat.fao.org/default.aspx; (c) Günther Fischer, Mahendra Shah, Francesco N. Tubiello, and Harrij Van Velhuizen, "Socio-Economic and Climate-Change Impacts on Agriculture: An Integrated Assessment, 1990–2080," *Philosophical Transactions of the Royal Society B*, vol. 360, no. 1463, November 29, 2005, 2067–83; (d) Günther Fischer, Mahendra Shah, and Harrij van Velhuizen, "Climate Change and Agricultural Vulnerability," International Institute for Applied Systems Analysis, World Summit on Sustainable Development, Johannesburg, 2002, 112-13, at http://www.iiasa.ac.at./Research/LUC/JB-Report.pdf; (e) David Grigg, *The World Food Problem* (Hoboken/London: Wiley Blackwell, 1994); (f) Nakicenovic and Swart (eds.), *IPCC Special Report on Emissions Scenarios*, 2000, at http://www.ipcc.ch/ipccreports/sres/emission/index.htm; (g) *World Food Summit: Technical Background Documents, Docs.* 1–15, UN Food and Agriculture Organization, 1996, at http://www.fao.org/wfs/ index_en.htm. Web sites were last accessed on July 29, 2009.

Lomborg cited the FAOSTAT URL without specifying the page or portion that supports his assertion; I could find nothing on the Web site in support of his claim. Likewise, Lomborg referenced page 2080 of the 2005 paper cited above from the *Philosophical Transactions of the Royal Society B*, but the specified page is focused exclusively on IPCC SRES scenario projections of undernourished populations; it does not support Lomborg's assertions pertaining to improvements over the past four decades.

Lomborg then cited two pages from a 2002 report, but the pages specified (pp. 112–13) also focused exclusively on projected impacts of climate change on hunger, and in fact fell within a section of the report titled "Impact of Climate Change on the Number of People at Risk of Hunger." These pages did not support Lomborg's assertion.

Lomborg also cited a 1994 book, *The World Food Problem*, without specifying any page or pages in the book to support his assertion. Lomborg likewise cited the 2000 IPCC *Special Report on Emissions Scenarios*, without specifying any pages that would support his assertion. Lomborg also cited Technical Background Documents issued by the FAO without specifying which parts of the documents might support his assertion.

12. Lomborg, *Cool It*, 103.
13. IPCC Fourth Assessment Report (2007): Working Group II: Impacts, Adaptation, and Vulnerability: Chapter 5, "Food, Fibre, and Forest Products," 275, last accessed on July 29, 2009, at http://www.ipcc.ch/pdf/assessment-report/ar4/wg2/ar4-wg2-chapter5.pdf.
14. Ibid., 298.
15. IPCC Fourth Assessment Report (2007): Working Group II: Impacts, Adaptation, and Vulnerability, "Technical Summary," 47, last accessed on July 29, 2009, at http://www.ipcc.ch/pdf/assessment-report/ar4/wg2/ar4-wg2-ts.pdf.
16. Ibid., 48.
17. Lomborg, *Cool It*, 103.
18. Ibid., 190.
19. Ibid., 103.
20. M. L. Parry, C. Rosenzweig, A. Iglesias, et al., "Effects of Climate Change on Global Food Production Under SRES Emissions and Socio-Economic Scenarios," *Global Environmental Change*, vol. 14, no. 1, April 2004, 53.
21. Ibid.
22. Lomborg, *Cool It*, 190.
23. G. Fischer, M. Shah, F. N. Tubiello, and H. van Velhuizen, "Socio-Economic and Climate Change Impacts on Agriculture: An Integrated Assessment, 1990–2080," *Philosophical Transactions of the Royal Society B*, vol. 360, no. 1463, November 29, 2005, 2067.
24. Ibid., 2079–80.
25. Lomborg, *Cool It*, 190.
26. Günther Fischer, Mahendra Shah, and Harrij van Velhuizen, "Climate Change and Agricultural Vulnerability," A Special Report Prepared by the International Institute for Applied Systems Analysis Under United Nations Institutional Contract Agreement No. 1113 on "Climate Change and Agricultural Vulnerability" as a Contribution to the World Summit on Sustainable Development, Johannesburg 2002, 120, 121, last accessed on July 29, 2009, at http://www.iiasa.ac.at/Research/LUC/JB-Report.pdf.
27. Günther Fischer, Harrij van Velthuizen, Mahendra Shah, and Freddy O. Nachtergaele, "Global Agro-ecological Assessment for Agriculture in the 21st Century: Methodology and Results," International Institute for Applied Systems Analysis, Laxenburg, Austria and the Food and Agriculture

Organization of the United Nations, Viale delle Terme di Caracalla, Rome, Italy, 2002, 105, last accessed on July 29, 2009, at http://www.iiasa.ac.at/Admin/PUB/Documents/RR-02-002.pdf.

28. Martin Parry, Cynthia Rosenzweig, and Matthew Livermore, "Climate Change, Global Food Supply and Risk of Hunger," *Philosophical Transactions of the Royal Society B*, vol. 360, no. 1463, October 24, 2005, 2134.
29. Cynthia Rosenzweig and Martin L. Parry, "Potential Impact of Climate Change on World Food Supply," *Nature*, vol. 367, January 13, 1994, 133.
30. Lomborg, *Cool It*, 103.
31. Ibid., 103–5.
32. The IPCC Fourth Assessment Report (2007): Working Group II: Impacts, Adaptation, and Vulnerability, "Technical Summary," 43.
33. Ibid., 48.
34. The IPCC Fourth Assessment Report (2007): Working Group II: Impacts, Adaptation, and Vulnerability: Chapter 10, "Asia," 487, last accessed on July 29, 2009, at http://www.ipcc.ch/pdf/assessment-report/ar4/wg2/ar4-wg2-chapter10.pdf.
35. The IPCC Fourth Assessment Report (2007): Working Group II: Impacts, Adaptation, and Vulnerability, "Summary for Policymakers," 12, 13, 16, 18, last accessed on July 29, 2009, at http://www.ipcc.ch/pdf/assessment-report/ar4/wg2/ar4-wg2-spm.pdf; "Technical Summary," 43, 47, 48, 68, 75; Chapter 8, "Human Health," 399, 401, 406, 407, 413, last accessed on July 29, 2009, at http://www.ipcc.ch/pdf/assessment-report/ar4/wg2/ar4-wg2-chapter8.pdf; Chapter 9, "Africa," 446–47, 458, last accessed on July 29, 2009, at http://www.ipcc.ch/pdf/assessment-report/ar4/wg2/ar4-wg2-chapter9.pdf; Chapter 10, "Asia," 482.

Chapter 12. On Water Shortages

1. Bjørn Lomborg, *Cool It: The Skeptical Environmentalist's Guide to Global Warming* (New York: Knopf, 2007), 108. To read Lomborg's claims that global warming will reduce the number of people living in water-stressed areas, see *Cool It*, 109-10.
2. IPCC Fourth Assessment Report (2007): Working Group II: Impacts, Adaptation, and Vulnerability: Chapter 3, "Freshwater Resources and Their Management," 178–79, last accessed on July 27, 2009, at http://www.ipcc.ch/pdf/assessment-report/ar4/wg2/ar4-wg2-chapter3.pdf.
3. Ibid.

4. Lomborg, *Cool It*, 108, 192, 239.
5. UNESCO, *Water: A Shared Responsibility: The United Nations World Water Development, Report* 2, Chapter 2: "The Challenges of Water Governance," 2006, 45, last accessed on July 27, 2009, at http://www.unesco.org/water/wwap/wwdr/wwdr2/pdf/wwdr2_ch_2.pdf.
6. UNESCO, *Water: A Shared Responsibility: The United Nations World Water Development, Report* 2, 2006, Section 1: "Changing Contexts: Index of Non-Sustainable Use," 2, last accessed on July 27, 2009, at http://www.unesco.org/water/wwap/wwdr/wwdr2/pdf/wwdr2_section_1.pdf.
7. Lomborg, *Cool It*, 192.
8. This is according to a search of the document, UNESCO, *Water: A Shared Responsibility: The United Nations World Water Development, Report* 2, last accessed on October 4, 2009, at http://unesdoc.unesco.org/images/0014/001454/145405E.org.
9. Lomborg, *Cool It*, 111.
10. Ibid., 192. Lomborg also wrote in the same reference: "Compare this to about $100 billion over the period in (Rijsberman, 2004:521)." According to this estimate, the cost per year would be more than $10 billion, not $4 billion.
11. Ibid.
12. Jérémie Toubkiss, "Costing MDG Target 10 on Water Supply and Sanitation: Comparative Analysis, Obstacles and Recommendations," World Water Council, 2006, last accessed on July 27, 2009, at http://www.worldwatercouncil.org/fileadmin/wwc/Library/Publications_and_reports/FullTextCover_MDG.pdf.
13. Ibid., 7.
14. Lomborg, *Cool It*, 111–12.
15. Ibid.

Chapter 13. Lomborg's Triple-A Rating

1. "Principles Governing IPCC Work: Approved at the Fourteenth Session (Vienna, 1–3 October 1998) on 1 October 1998, amended at the 21st Session (Vienna, 3 and 6–7 November 2003) and at the 25th Session (Mauritius, 26–28 April 2006)," last accessed on July 29, 2009, at http://www.ipcc.ch/pdf/ipcc-principles/ipcc-principles.pdf.
2. "A Calm Voice in a Heated Debate," *Wall Street Journal*, September 13, 2007; "Challenges to Both Left and Right in U.S. on Global Warming," *New York Times*, November 13, 2007.
3. "A Calm Voice in a Heated Debate."

4. "Chill Out: Stop Fighting over Global Warming—Here's the Smart Way to Attack It," *Washington Post*, October 7, 2007.
5. "Challenges to Both Left and Right in U.S. on Global Warming."
6. Arctic Climate Impact Assessment, 2004, 509, last accessed on July 29, 2009, at http://www.acia.uaf.edu/pages/scientific.html.
7. Ian Stirling, "Ice Bear: Icon of the Arctic" in *Planet Ice: A Climate for Change* (Seattle, Washington: Braided River, 2009).
8. Tierney, "'Feel Good' vs. 'Do Good,'" *New York Times*, September 11, 2007.
9. IPCC Fourth Assessment Report (2007): Working Group II: Impacts, Adaptation, and Vulnerability: Chapter 8, "Human Health," 396–405, last accessed on July 29, 2009, at http://www.ipcc.ch/pdf/assessment-report/ar4/wg2/ar4-wg2-chapter8.pdf.
10. Ibid., 393.
11. See "About Foreign Affairs: Staff," last accessed on July 29, 2009, at http://www.foreignaffairs.org/about/staff.
12. Cooper, "Review: *Cool It: The Skeptical Environmentalist's Guide to Global Warming*," *Foreign Affairs*, January/February 2008, last accessed on July 29, 2009, at http://www.foreignaffairs.org/20080101fashortreview87119/bj-rn-lomborg/cool-it-the-skeptical-environmentalist-s-guide-to-global-warming.html.
13. United Nations Framework Convention on Climate Change, 1992, article 2, last accessed on July 29, 2009, at http://unfccc.int/resource/docs/convkp/conveng.pdf.
14. John Tierney, "Honest Science in Washington," *New York Times*, TierneyLab Blog, February 23, 2009, last accessed on July 29, 2009, at http://tierneylab.blogs.nytimes.com/2009/02/23/honest-science-in-washington/?scp=4&sq=Holdren%20Pielke&st=cse.
15. Roger A. Pielke Jr., *The Honest Broker: Making Sense of Science in Policy and Politics* (Cambridge, U.K.: Cambridge University Press, 2007).
16. See Pielke Jr., *The Honest Broker: Making Sense of Science in Policy and Politics*, pp. 128–29, 132.
17. Tierney, "Honest Science in Washington."
18. Pielke, *The Honest Broker*, 121, 124.

Chapter 14. How Wrong Was Lomborg?

1. "Secretary Kempthorne Announces Decision to Protect Polar Bears Under Endangered Species Act," U.S. Department of the Interior, Office of

the Secretary, News Release, May 12, 2008, last accessed on July 29, 2009, at http://www.fws.gov/home/feature/2008/polarbear012308/pdf/DOI_polar_bears_news_release.pdf.

2. "Polar Bears at Risk of Drowning in the Chukchi Sea," Center for Biological Diversity (via Common Dreams), August 21, 2008, last accessed on July 29, 2009, at http://www.commondreams.org/news2008/0821-06.htm.
3. Ibid.
4. Bjørn Lomborg, *Cool It: The Skeptical Environmentalist's Guide to Global Warming* (New York: Knopf, 2007), 6.
5. "Arctic Summers Ice-Free 'by 2013,'" BBC News, December 12, 2007, last accessed on July 29, 2009, at http://news.bbc.co.uk/2/hi/science/nature/7139797.stm.
6. "Thick, Old Ice Nearly Gone," *Discovery News*, January 18, 2008, last accessed on July 29, 2009, at http://dsc.discovery.com/news/2008/01/18/arctic-ice-melt.html.
7. "Arctic Ice Continues to Thin," *New Scientist*, August 2, 2008, last accessed on July 29, 2009, at http://www.newscientist.com/article/mg19926673.400.
8. "Canadian Scientists Predict Ice-Free Arctic Summer in 6 yrs," Australian Broadcasting Corporation, December 14, 2008, last accessed on July 29, 2009, at http://www.abc.net.au/news/stories/2008/12/14/2445828.htm.
9. Ian Stirling and Andrew E. Derocher, "Melting Under Pressure: The Real Scoop on Climate Warming and Polar Bears," *The Wildlife Professional*, Fall 2007, 24, 43, last accessed on July 29, 2009, at http://www.biology.ualberta.ca/faculty/andrew_derocher/uploads/abstracts/Stirling_Derocher_Wildlife_Professional_PB_climate_2007.pdf.
10. Lomborg, *Cool It*, 53, 59.
11. "Glaciers Melt 'At Fastest Rate in Past 5,000 Years,'" *Guardian*, March 16, 2008, last accessed on July 29, 2009, at http://www.guardian.co.uk/environment/2008/mar/16/glaciers.climatechange1.
12. "Melting Mountains a 'Time Bomb' for Water Shortages," Reuters, April 15, 2008, last accessed on July 29, 2009, at http://www.reuters.com/article/environmentNews/idUSL145733352008O415?sp=true.
13. "Tibetan Glaciers Melting at Stunning Rate," *Discovery News*, November 24, 2008, last accessed on July 29, 2009, at http://dsc.discovery.com/news/2008/11/24/tibet-glaciers-warming.html.
14. Eric J. Steig, David P. Schneider, Scott D. Rutherford, et al., "Warming

of the Antarctic Ice-Sheet Surface Since the 1957 International Geophysical Year," *Nature*, vol. 457, January 22, 2009, 459; "Study Finds New Evidence of Warming in Antarctica," *New York Times*, January 22, 2009.

15. Eric Rignot, Jonathan L. Bamber, Michael R. van den Broeke, et al., "Recent Antarctic Ice Mass Loss from Radar Interferometry and Regional Climate Modeling," *Nature Geoscience*, vol. 1, February 2008, 106.
16. "Antarctic Glaciers Melting More Quickly," *San Francisco Chronicle*, January 26, 2008, last accessed on July 29, 2009, at http://www.sfgate.com/cgi-bin/article.cgi?f=/c/a/2008/01/26/MN50UM20C.DTL.
17. "Gigantic Antarctic Ice Chunk Collapses," Associated Press (via *Discovery News*), March 25, 2008, last accessed on July 29, 2009, at http://dsc.discovery.com/news/2008/03/25/antarctica-ice-collapse.html.
18. "Antarctic Ice Shelf 'Hanging on by a Thread,'" ABC News (Australian Broadcasting Corporation), July 11, 2008, last accessed on July 29, 2009, at http://www.abc.net.au/news/stories/2008/07/11/2301297.htm.
19. "Antarctic Ice Shelf Set to Collapse Due to Warming," Reuters, January 19, 2009, last accessed on July 29, 2009, at http://www.reuters.com/article/environmentNews/idUSTRE50I4G520090120?pageNumber=1&virtualBrandChannel=0.
20. "Forecast for Big Sea Level Rise," BBC News, April 15, 2008, last accessed on July 29, 2009, at http://news.bbc.co.uk/2/hi/science/nature/7349236.stm.
21. "Sea Level Could Rise by 150cm, US Scientists Warn," *Guardian*, December 16, 2008, last accessed on July 29, 2009, at http://www.guardian.co.uk/environment/2008/dec/16/climatechange-scienceofclimatechange.
22. "Ocean Expected to Rise 5 Feet Along Coastline," *San Francisco Chronicle*, March 12, 2009, last accessed on July 29, 2009, at http://www.sfgate.com/cgi-bin/article.cgi?f=/c/a/2009/03/11/MNTK16DEBF.DTL.
23. "Scientists to Issue Stark Warning over Dramatic New Sea Level Figures," *Guardian*, March 8, 2009, last accessed on July 29, 2009, at http://www.guardian.co.uk/science/2009/mar/08/climate-change-flooding.
24. "Northeast US to Suffer Most from Future Sea Rise," Associated Press (via *USA Today*), March 16, 2009, last accessed on July 29, 2009, at http://www.usatoday.com/weather/climate/globalwarming/2009-03-15-sea-level-rise-northeast_N.htm.
25. Lomborg, *Cool It*, 60.
26. Ibid., 62–63.
27. Ibid., 103.

28. "World Must Double Food Production by 2050: FAO Chief," Agence France Presse (via Common Dreams), January 26, 2009, last accessed on July 29, 2009, at http://www.commondreams.org/headline/2009/01/26-8.
29. "Emperor Penguin 'Marching to Extinction by End of the Century,'" *Independent*, January 27, 2009, last accessed on July 29, 2009, at http://www.independent.co.uk/environment/nature/emperor-penguin-marching-to-extinction-by-end-of-the-century-1516804.html.
30. "Carbon Dioxide Levels Rising Faster Than Predicted, Study Says," Bloomberg, October 22, 2007, last accessed on July 29, 2009, at http://www.bloomberg.com/apps/news?pid=20601081&sid=apGVMyvFvMfY.
31. "Carbon Dioxide at Record High, Stoking Warming: WMO," Reuters, November 23, 2007, last accessed on July 29, 2009, at http://www.reuters.com/article/environmentNews/idUSFLE36103820071125.
32. Global Carbon Project, "Growth in the Global Carbon Budget: Updated Global Carbon Budget Released," September 25, 2008, last accessed on July 29, 2009, at http://www.globalcarbonproject.org/global/pdf/Press%20Release_GCP.pdf.
33. Ibid.
34. "Carbon Cuts 'Only Give 50/50 Chance of Saving Planet,'" *Independent*, March 9, 2009, last accessed on July 29, 2009, at http://www.independent.co.uk/environment/climate-change/carbon-cuts-only-give-5050-chance-of-saving-planet-1640154.html.
35. "Global Warming 'Irreversible' for Next 1000 Years: Study," Agence France Presse, January 27, 2009, last accessed on July 29, 2009, at http://www.google.com/hostednews/afp/article/ALeqM5ghGvjbYscejTKon1TkD_K4E3t_lw.
36. "Report: Some Climate Damage Already Irreversible," Associated Press (via *USA Today*), January 26, 2009, last accessed on July 29, 2009, at http://www.usatoday.com/weather/climate/globalwarming/2009-01-26-climate-change-irreversible_N.htm.
37. James Hansen, Sato Makiko, Pushker Kharecha, et al., "Target Atmospheric CO2: Where Should Humanity Aim?" *The Open Atmospheric Science Journal*, vol. 2, December 2008, last accessed on July 29, 2009, at http://www.columbia.edu/jeh1/2008/TargetCO2_20080407.pdf.
38. "Gore to UN: 350 or Bust: Al Gore Calls for 350 ppm Goal at Poznan Climate Summit," Grist, posted December 12, 2008, last accessed on July 29, 2009, at http://gristmill.grist.org/story/2008/12/12/123136/53.
39. "Heat and Hope: Time Running Out for Steep Emissions Cuts," World-

watch Institute, press release, January 13, 2009, last accessed on July 29, 2009, at http://www.worldwatch.org/node/5987.

40. Worldwatch Institute, *State of the World 2009: Into a Warming Word* (New York: W.W. Norton, 2009).
41. Nicholas Stern, *The Economics of Climate Change: The Stern Review* (Cambridge, U.K.: Cambridge University Press, 2007).
42. "Lord Stern on Global Warming: It's Even Worse Than I Thought," *Independent*, March 13, 2009, last accessed on July 29, 2009, at http://www.independent.co.uk/environment/climate-change/lord-stern-on-global-warming-its-even-worse-than-i-thought-1643957.html.
43. "US Still Skeptical About Global Warming: Survey," Agence France Presse (via Common Dreams), March 13, 2009, last accessed on July 29, 2009, at http://www.commondreams.org/headline/2009/03/13-3.
44. Al Gore, *The Assault on Reason* (New York: Penguin, 2007).

INDEX